PERIODICALS 058-457

Global Issues Series

General Editor: **Jim Whitman**

This exciting new series encompasses three principal themes: the interaction of human and natural systems; cooperation and conflict; and the enactment of values. The series as a whole places an emphasis on the examination of complex systems and causal relations in political decision-making; problems of knowledge; authority, control and accountability in issues of scale; and the reconciliation of conflicting values and competing claims. Throughout the series the concentration is on an integration of existing disciplines towards the clarification of political possibility as well as impending crises.

Titles include:

Brendan Gleeson and Nicholas Low (*editors*)
GOVERNING FOR THE ENVIRONMENT
Global Problems, Ethics and Democracy

Roger Jeffery and Bhaskar Vira (*editors*)
CONFLICT AND COOPERATION IN PARTICIPATORY NATURAL RESOURCE
MANAGEMENT

Ho-Won Jeong (*editor*)
GLOBAL ENVIRONMENTAL POLICIES
Institutions and Procedures

W. Andy Knight
A CHANGING UNITED NATIONS
Multilateral Evolution and the Quest for Global Governance

W. Andy Knight
ADAPTING THE UNITED NATIONS TO A POSTMODERN ERA
Lessons Learned

Graham S. Pearson
THE UNSCOM SAGA
Chemical and Biological Weapons Non-Proliferation

Andrew T. Price-Smith (*editor*)
PLAGUES AND POLITICS
Infectious Disease and International Policy

Michael Pugh (*editor*)
REGENERATION OF WAR-TORN SOCIETIES

Bhaskar Vira and Roger Jeffery (*editors*)
ANALYTICAL ISSUES IN PARTICIPATORY NATURAL RESOURCE MANAGEMENT

Simon M. Whitby
BIOLOGICAL WARFARE AGAINST CROPS

Global Issues Series
Series Standing Order ISBN 0–333–79483–4
(*outside North America only*)

You can receive future titles in this series as they are published by placing a standing order. Please contact your bookseller or, in case of difficulty, write to us at the address below with your name and address, the title of the series and the ISBN quoted above.

Customer Services Department, Macmillan Distribution Ltd, Houndmills, Basingstoke, Hampshire RG21 6XS, England

Conflict and Cooperation in Participatory Natural Resource Management

Edited by

Roger Jeffery
Professor of Sociology of South Asia
University of Edinburgh

and

Bhaskar Vira
University Assistant Lecturer in Environment and Development
Department of Geography
University of Cambridge

First published 2001 by
PALGRAVE
Houndmills, Basingstoke, Hampshire RG21 6XS and
175 Fifth Avenue, New York, N. Y. 10010
Companies and representatives throughout the world

PALGRAVE is the new global academic imprint of
St. Martin's Press LLC Scholarly and Reference Division and
Palgrave Publishers Ltd (formerly Macmillan Press Ltd).

ISBN 0–333–79277–7

This book is printed on paper suitable for recycling and
made from fully managed and sustained forest sources.

A catalogue record for this book is available
from the British Library.

Library of Congress Cataloging-in-Publication Data
Conflict and cooperation in participatory natural resource
management / edited by Roger Jeffery and Bhaskar Vira.
    p. cm. — (Global issues)
    Paper presented at a workshop held at Mansfield College,
Oxford, on April 6 and 7, 1998.
    Includes bibliographical references and index.
    ISBN 0–333–79277–7
    1. Natural resources—Management—Congresses.
    2. Environmental policy—International cooperation–
—Congresses. I. Jeffery, Roger. II. Vira, Bhaskar. III. Series.
HC85 .C66 2001
333.7—dc21
                                        2001021197

10 9 8 7 6 5 4 3 2 1
10 09 08 07 06 05 04 03 02 01

Printed in Great Britain by Antony Rowe Ltd, Chippenham, Wiltshire

Contents

List of Tables, Figures and Box   vii

Preface and Acknowledgements   ix

Notes on the Contributors   xi

1   Introduction   1
    Roger Jeffery and Bhaskar Vira

**Part I   Where Local Conflicts over Resource Use
Make Participation Unlikely**   17

2   Conflict Management and Mobility among
    Pastoralists in Karamoja, Uganda   19
    Maryam Niamir-Fuller

3   The Social Context of Environmental Education:
    The Case of the Amboseli Ecosystem, Kajiado, Kenya   39
    Ngeta Kabiri

**Part II   Local-Level Projects Attempting to
Overcome Unsupportive National Contexts**   61

4   Addressing Livelihood Issues in Conservation-
    oriented Projects: A Case Study of Pulicat Lake,
    Tamil Nadu, India   63
    Devaki Panini

5   Land Husbandry for Sustainable Agricultural
    Development in a Subsistence Farming Area of Malawi:
    Farmer Adoption of Introduced Techniques   75
    Max Kelly

**Part III   Learning from Success: Where National
Policies are Supportive, but Participative
Action is Novel**   97

6   Local Management of Sahelian Forests   99
    Paul Kerkhof

 7   CAMPFIRE: Tonga Cosmovision and
     Indigenous Knowledge   113
     Backson Sibanda

 8   Problems of Intra- and Inter-group Equity in Community
     Forestry: Evidence from the Terai Region of Nepal   129
     Rabindra Nath Chakraborty

 9   Benefits to Villagers in Maharashtra, India, from
     Conjunctive Use of Water Resources   150
     Frank Simpson and Girish Sohani

**Part IV   Learning from Success: Supportive
National Policies and Local Initiatives   169**

10   Creating New Knowledge for Soil and Water
     Conservation in Bolivia   171
     Anna Lawrence

11   Devolution of Decision-making: Lessons from
     Community Forest Management at the Kilum–Ijim
     Forest Project, Cameroon   189
     David Thomas, Anne Gardner and John DeMarco

12   Changing Natural Resource Research and Development
     Capability: Whither Social Capital?   204
     Stephen Biggs, Harriet Matsaert and Adrienne Martin

References   227

Index   241

List of Tables, Figures and Box

Tables

1.1	Typology of participation	3
5.1	Research sites within Salima ADD	79
5.2	Average land holdings within the study area and percentage of smallholders farming less than 1 hectare	80
5.3	Ranking of problems in each village by the village members	85
5.4	The total number of households who feel they actively participate in the PROSCARP project	86
5.5	Numbers of marker ridges constructed by village	88
5.6	Average distance walked for the collection of firewood	91
5.7	Percentage of covered and uncovered pit latrines in the villages	92
5.8	Results of a survey of farmer opinion towards the PROSCARP project	93
7.1	How is traditional knowledge utilised in CAMPFIRE (1)?	124
7.2	How is traditional knowledge utilised in CAMPFIRE (2)?	124
8.1	Community forests transferred, Banke District	134
8.2	Registered forest user groups in Dhanusha District	136
10.1	Rural people's practices for SWC in the temperate valleys	179

Figures

9.1	Location of Akole Taluka, Ahmednagar District, Maharashtra State, India	151
9.2	Location of study area in Akole Taluka	153
10.1	Stages of information flow in participatory research	175

11.1 Summary of conditions for devolving rights for
natural resource management to local communities   196

Box

10.1 The process used to help farmers plan experiments
during the participatory workshops   182

Preface and Acknowledgements

Attempts to manage natural resources through collaboration rather than competition, through negotiation rather than by fiat, by agreement rather than conflict, have become a touchstone for many who see these efforts as the harbinger of global sustainable development. But thoroughgoing, independent assessment of the successes and failures of experimental programmes around the world, and the conditions within which such efforts might be made with some chances of success, are still in their infancy. In April 1998 over 120 academics and consultants, with representatives of NGOs and governments, came together to discuss papers that addressed these themes. It was an enjoyable three days, with considerable crossing over of the boundaries sometimes found between people from these different constituencies. Debate and discussion went on well into the night, generating ideas for new research, for changed practices within donor agencies and for new ways to thinks about some of the key issues.

This book and its partner (Bhaskar Vira and Roger Jeffery, 2001, eds, *Analytical Issues in Participatory Natural Resource Management*) bring together some of the papers presented at the Oxford conference. We have attempted to convey something of the excitement generated there, by making these papers more widely available. Inevitably, in the process of revising, editing and updating and preparing for publication, some of the papers have been amended considerably, in part to take account of changes since 1998 in the political, economic and social circumstances in the many different parts of the world considered here. In some cases (such as in Indonesia) the political changes have been so considerable that predictions based on material before 1999 offer a poor guide to opportunities in the future. None the less, we think these papers provide valuable insights into activities, structures and processes with great relevance for the twenty-first century.

We are grateful to many people for their work and support in making the conference possible and helping to see the project through to publication. The Economic and Social Research Council's grant funded the conference by supporting the administrative overheads and allowing us to invite scholars and consultants from India, Africa and Latin America. Alistair Scott, in the office of the Global Environmental Change Initiative in Sussex was particularly supportive of our efforts. Much of the preparatory work for the conference was done by Esther Dermott, and liaison with Mansfield College was sensitively handled by Anne Maclachlan. Keynote talks given at the conference by Melissa Leach, Iain Scoones and Ian Swingland helped to get the conference off on a sound footing. Discussants were Darrell Posey (Mansfield College), Catherine Locke (University of East Anglia), Philip Woodhouse (University of Manchester) and Helle Qwist-Hoffman (FAO). Jim Whitman encouraged us to submit a proposal for publication to Palgrave as part of the Global Issues Series, and Karen Brazier and Eleanor Birne coordinated the project for the publishers. Valuable assistance with editing the papers was provided by Lucy Welford in Cambridge and Colin Millard in Edinburgh. The Centre for South Asian Studies at the University of Edinburgh provided financial support for the editing. None of the above is responsible for the views expressed here, nor for any errors and omissions that remain.

As with all such activities, the families of the editors have suffered in various ways: we are grateful for their tolerance and aware of our debts to Shiraz and Kartik (BV) and Tricia, Laura and Kirin (RJ).

Notes on the Contributors

Stephen Biggs is a senior lecturer in the School of Development Studies at the University of East Anglia. His interests are in rural development, natural resources and agricultural policy, research policy and management, NGOs, participatory technology development, capability and institutional development, research methods, and approaches to planning, monitoring and evaluation. His experiences are mainly in Asia and Southern Africa. Recent publications include: S. Biggs and G. Smith, 1998, 'Beyond Methodologies: Coalition-Building for Participatory Technology Development', *World Development*, 26, 2: 239–48; and S. Biggs and H. Matsaert, 1999, 'An actor oriented approach for strengthening research and development capabilities in natural resources systems', *Public Administration and Development*, no. 9.

Rabindra Nath Chakraborty studied mechanical engineering and economics at the Technical University of Darmstadt, Germany. From 1990 to 1995, he was a research staff member at the Department of Law and Economics of the Technical University of Darmstadt. Since 1995, he has been at the German Development Institute in Berlin. Local institutions of natural resource management in South Asia have been one focus of his research work; another has been the macroeconomic impact of environmental degradation on growth and income distribution in developing countries. Recent publications include: 'Linkages between income distribution and environmental degradation in rural India', in Stig Toft Madsen (ed.), 1999, *State, Society and the Environment in South Asia*, 165–99; and *Growth, Environmental Degradation, and Income Distribution in Developing Countries. The Case of India* (in German with English summary).

John DeMarco is a specialist in rural development. From 1983 to 1988 he was country representative and project manager for the Institute for the Study and Application of Integrated Development (ISAID) in Niger. From 1990 to 1991 he worked as a consultant on a variety of projects, including a rural water supply project in Swaziland. From 1993 to 1994 he was director for a CARE rural development project in Cameroon. The 'Projet Stratégie d'Occupation des Sols – Louti Nord' aimed to contribute to more sustainable use of natural resources by providing technical assistance and support of community initiatives. He joined BirdLife

International in 1995, as project manager of the Ijim Mountain Forest Project (now Kilum–Ijim Forest Project).

Anne Gardner is a specialist in rural development planning. From 1984 to 1988 she was country representative and project manager for the Institute for the Study and Application of Integrated Development (ISAID) in Niger. She was jointly responsible for the design and development of the Programme d'appui aux actions villageoises, a rural development project providing organisational, technical and material assistance to villages. In 1991 she joined Oxfam-Canada as administrator of their Horn of Africa Programme. From 1993 to 1994 she was planning and research adviser to the 'Projet Stratégie d'Occupation des Sols – Louti Nord', a CARE project in northern Cameroon. She joined BirdLife International as project manager of the Ijim Mountain Forest Project (now Kilum–Ijim Forest Project) in 1995.

Roger Jeffery has been Professor of Sociology of South Asia at the University of Edinburgh since 1997. His research in India since 1971 has covered health policy-making as well as village-based social demographic fieldwork in north India. He was a principal investigator for an ESRC-funded project looking at Joint Forest Management in four Indian States (1994–97). Recent publications include: R. Jeffery (ed.), 1998, *The Social Construction of Indian Forests*; and R. Jeffery and N. Sundar (eds), 1999, *A New Moral Economy for India's Forests?*

Ngeta Kabiri holds a BA from Kenyatta University and MA degrees from Kenyatta and Yale Universities. He has taught at Kenyatta University (Kenya) and at Prairie View A&M University (USA). He is currently a PhD student in the Political Science Department, University of North Carolina at Chapel Hill. His research interests are in the politics of the environment in Africa, and he has done research among the Maasai on agropastoralism and fertility regulation. He has presented conference papers on the environmental factor in theories of fertility regulation in Africa.

Max Kelly is a PhD student in the School of Geography at Kingston University working on sustainable agriculture and livelihood issues in Malawi. After taking a B.Ag.Sc at University College Dublin she worked on soil erosion problems in Australia for a short time. An MSc in Resource Management at Aberdeen University led to the opportunity to follow up her interest in soil conservation through the research in Malawi reported here.

Paul Kerkhof is a research and development worker specialising in forest resources in sub-Saharan Africa, with an emphasis on East Africa and the Sahel. He has worked for a dozen bilateral and private organisations in 14 African countries, and has been employed since 1996 by SOS Sahel (UK) as a researcher. Having worked for different organisations in Kenya, he has witnessed the competitive atmosphere which may exist in development, and decided to try to break communication barriers by writing a book about the experiences of different organisations. This was published by the Panos Institute, London, in 1990, entitled *Agroforestry in Africa. A Survey of Project Experience.*

Anna Lawrence has a background in plant ecology and forestry, and worked in agroforestry research and extension in Bolivia. Through ethnobotanical work and social research she has become particularly interested in the interactions between people and natural resources, and how different perceptions and values of species and ecosystems relate to understanding and management of biodiversity. With collaborators in South America and Asia she has used participatory methods for facilitating communication about these issues between different stakeholders, and for involving local people in decision-making. The work reported in this chapter was conducted while she was a Research Fellow in the Agricultural Extension and Rural Development Department, University of Reading. She is now a Senior Research Associate at the Centre for Natural Resources and Development, Green College, University of Oxford. Recent publications include 'Going with the flow or an uphill struggle? Directions for participatory research in hillside environments', *Mountain Research and Development* 19,3 (1999), 203–12; and A. Lawrence, J.J.F. Barr and G.S. Haylor, *Stakeholder Approaches to Planning Participatory Research.* ODI Agricultural Research and Extension Network Paper no. 91. Overseas Development Institute, London, 1999.

Adrienne Martin is a member of the Social Development Group, Natural Resource Institute, University of Greenwich, Chatham. Her interests are in development and training in participatory research and its institutionalisation within agricultural research and extension systems; analysis of rural and urban households' strategies for livelihood security; gender and technology development and indigenous systems. Her experience has been mainly in Africa and the Middle East. Recent publications include: P. Goldey, S. Le Breton, A. Martin and R. Marcus, 'Approaches to address gender specific needs in relation to access to technological change', *Agricultural Systems* 55, 2 (1997), 155–72; and A. Martin and J. Sherington, 'Participatory methods – implementation, effectiveness, and institutional context', *Agricultural Systems* 55,2 (1997), 195–216.

Harriet Matsaert is a social anthropologist currently based in Harare. She has worked for the Natural Resources Institute and DFID on Farming Systems Research and Extension teams in the Ministry of Agriculture in Kenya and Namibia. Particular interests include participatory technology development working with farmers, local artisans, academic and private sectors, and institutional capacity-building for agricultural research and development. Publications include: Matsaert, Mellis and Mwaniki, 'Tillage research challenges tool makers in Kenya', in *Farmers' Research in Practice*, 1997, and Matsaert, Gibbon, Mutwamwezi and Kakukuru, 'Heterogeneity and multiple realities – the Kavango farming systems team's experiences of understanding and working with difference'. Paper prepared for International Farming Systems Research and Extension Workshop, Pretoria, November 1998.

Maryam Niamir-Fuller holds a doctorate in range management, and has specialised in pastoral development for the past 20 years. Her work on pastoralism includes research (Dinka of Sudan, Barabaig of Tanzania), development project leadership (Senegal) and numerous consultancies for project development and evaluation for multilateral, bilateral and NGO donors in Africa. She has organised four International Technical Consultations on Pastoral Development for UNSO. She is currently engaged as a land degradation adviser and task manager for the development of biodiversity conservation projects for GEF/UNDP. Her most recent publication is as editor and contributor to *Managing Mobility in African Rangelands: The Legitimization of Transhumance* (1999).

Devaki Panini was trained as a lawyer and specialises in environmental law. She has worked as programme officer and law officer with WWF-India in its Wetlands Division and Centre for Environmental Law. While working in the Wetlands Division, she coordinated an ecological restoration project that aimed to restore Pulicat Lake – a threatened wetland system in Tamil Nadu – with the participation of the fishermen of the lake. Her chapter in this book is based on insights and experience drawn during her field visits to Pulicat Lake from 1996 to 1998. Devaki is presently based in New Delhi and is working as an independent legal consultant on environmental and public interest cases and legal research on environmental issues in India.

Backson Sibanda is Chief of Evaluation at the United Nations Environment Programme (UNEP). He holds a doctorate in natural resources management from Rhodes, South Africa. He also holds a masters degree in planning from Nairobi, Kenya and a BSc in geography, London. He has worked for international NGOs, governments and donor agencies. He taught part-time at the University of Zimbabwe from 1983 until 1992. He is a founder member of the SAPES/SARIPS Masters policy programme and a guest lecturer on environmental policy. He has provided consultancy services since 1984 in areas planning, rural development, natural resources management, programme planning, and project design, monitoring and evaluation.

Frank Simpson is Professor of Geology at the University of Windsor, Canada.

Girish Sohani works for the BAIF Development Research Foundation, Pune, Maharashtra, India.

David Thomas has a background in both natural and social sciences. From 1988 to 1989 he was research assistant at University College, London, where he participated in a study of the socio-economic impacts of a drainage and irrigation project bordering the Ichkeul National Park in northern Tunisia. He then worked for the IUCN Wetlands Programme in Nigeria from 1989 to 1992, as technical consultant to the Hadejia-Nguru Wetlands Conservation Project. The aim of the project was to promote the sustainable development of the water and wildlife resources of the Hadejia-Jama'are River Basin. The task of the technical consultant was to identify, through a participatory approach, constraints to local production and sustainable resource use, and to explore solutions to producers' problems. From 1992 to 1995, while at the Department of Geography, Cambridge University, he studied the long-term impacts of dams on ecology, economy and society in northern Nigeria. In 1995 he moved to Edinburgh University as a research associate exploring the impact of tourist developments on coastal ecosystems in Belize. He joined the secretariat of BirdLife International in 1996 as programme and projects manager, and is responsible for all aspects of management and support of several integrated conservation and development projects in Africa and Asia.

Bhaskar Vira is University Assistant Lecturer in Environment and Development at the Department of Geography, and a Fellow of Fitzwilliam College, at the University of Cambridge. His research examines the political economy of environmental and natural resource management in the developing world, with a special focus on forestry issues in India. He is particularly interested in issues of governance and the interaction of multiple stakeholders (especially government agencies, NGOs and local groups) in the context of resource use and management. Previous publications include: 'Institutional pluralism in forestry: Considerations of analytical and operational tools', *Unasylva*, 49 (1998); and 'Implementing joint forest management in the field: Towards an understanding of the community–bureaucracy interface', in R. Jeffery and N. Sundar (eds.), 1999, *A New Moral Economy for India's Forests? Discourses of Community and Participation.*

1
Introduction

Roger Jeffery and Bhaskar Vira

1. Participation as orthodoxy

Community-based institutions are increasingly the focus of attempts to manage natural resources in developing countries. The new orthodoxy – for example, among multinational donors such as the World Bank and FAO, as well as bilateral donors and many governments – is that environmental deterioration can best be reversed through involving local people (either directly or through the involvement of NGOs) in partnerships with the state, transforming the common experience of conflict into cooperation. Where natural resources are in decline, social conflicts are often major contributory factors, whether through the disastrous effects of wars (civil and international) or through attempts to get private benefits from previously collectively owned or managed resources. One solution proposed is that local people should be involved from the beginning in the design, management, monitoring and (increasingly) evaluation of projects and programmes to reverse a history of decay and destruction. As with all social innovations, however, this new orthodoxy may be observed rhetorically, but with little commitment to practice.

The chapters in this book, all originally presented as papers at a conference held in Oxford in April 1998, report on ways in which local people have been involved in natural resource management, and describe experimental programmes with a variety of forms of local involvement.[1] Some of the chapters describe small local projects, with potential for extension to a larger scale (for example, Simpson and Sohani, Panini, Lawrence, Thomas *et al.* and Kelly). Others produce local-level data on national or regional programmes (for example, Chakraborty, Sibanda and Kerkhof). Two chapters describe environmental management in

East Africa in the absence of projects or programmes specifically designed to deal with the interests of nomadic people (Niamir-Fuller and Kabiri), while the chapter by Biggs, Martin and Matsaert discusses ways in which participatory elements entered into mainstream agricultural research, development and extension activities. Several chapters are concerned with forest management. Those by Chakraborty and Kerkhof deal with settings in which forest regeneration or protection is a concern for general environmental reasons as well as a means to enhance local livelihoods, whereas Thomas *et al.*, Sibanda and Kabiri discuss settings where specific wildlife is also to be protected, either as endangered species (Thomas) or for tourism (Sibanda and Kabiri). Other chapters are concerned with the management of soil and water conservation (Simpson and Sohani, Lawrence and Kelly) or fishing (Panini). The chapters discuss projects and settings in sub-Saharan Africa (Niamir-Fuller, Sibanda, Kabiri, Kerkhof, Kelly and Thomas *et al.*); South America (Lawrence) and South Asia (Chakraborty, Simpson and Sohani and Panini). The examples used by Biggs *et al.* are drawn from Africa and India.

In each case, formal statements can be found committing an organisation, ministry or donor to involving local people; in many of the examples, however, only some local people have been involved, for some purposes, at some stages, in some ways – but not others (see Table 1.1 for a typology of the kinds of participation derived from Pretty 1995). More generally, the chapters discuss situations in countries where political institutions may very inadequately integrate and articulate the needs and interests of some of the groups involved.

Since the mid-1990s, donor agencies – notably in the World Bank's Environment Department, and the UK's Department for International Development (DfID) – have attempted to appraise and review the extent of participation in their projects (World Bank 1995). Some donors now regularly attempt to ensure the participation of key stakeholders at each stage of project development – from project definition through appraisal, implementation, monitoring and review – and have used different strategies to achieve this end. Of the most popular ways to involve local, poor and relatively unorganised resource users, support for NGOs and the use of participatory rural appraisal techniques are perhaps the best known (Chambers 1997; Tropp 1999). These major shifts in the approach to managing natural resources that have taken place since 1985 have affected experiences in most parts of the developing world: the chapters in this volume provide snapshots of how far they have delivered their potential in some significant settings.

Table 1.1 Typology of participation

1. Passive participation	People participate by being told what is going to happen or has already happened. It is a unilateral announcement by an administration or project management without listening to people's responses. The information being shared belongs only to external professionals.
2. Participation in information giving	People participate by answering questions posed by extractive researchers using questionnaire surveys or similar approaches. People do not have the opportunity to influence proceedings, as the findings are neither shared nor checked for accuracy.
3. Participation by consultation	People participate by being consulted and external agents listen to views. These external agents define both problems and solutions, and may modify these in the light of people's responses. Such a consultative process does not concede any share in decision-making and professionals are under no obligation to take on people's views.
4. Participation for material incentives	People participate by providing resources, for example labour, in return for food, cash or other material incentives. Much on-farm research falls into this category, as farmers provide the fields but are not involved in experimentation or the process of learning. It is very common to see this called participation, yet people have no stake in prolonging activities when the incentives end.
5. Functional participation	People participate by forming groups to meet predetermined objectives related to the project, which can involve the development or promotion of externally initiated social organisation. Such involvement does not tend to be at the early stages of project cycles or planning but rather after major decisions have been made. These institutions tend to be dependent on external initiators and facilitators, but may become self-dependent.
6. Interactive participation	People participate in joint analysis, which leads to action plans and the formation of new local institutions or the strengthening of existing ones. It tends to involve interdisciplinary methodologies that seek multiple perspectives, and make use of systematic and structured learning processes. These groups take control over local decisions and so people have a stake in maintaining structures or practices.
7. Self-mobilisation	People participate by taking initiatives independent of external institutions to change systems. They develop contacts with external institutions for resources and technical advice that they need, but retain control over how resources are used. Such self-initiated mobilisation and collective action may or may not challenge existing inequitable distributions of wealth or power.

Source: Derived by Max Kelly from Pretty (1995).

2. Common themes

In introducing these case studies of the success and failure of participatory management of natural resources, we want to stress the following issues: first, we consider the basis of the participation and the significance to projects and programmes of the resources brought by local people to the negotiation of the form of the agreements; second, we analyse the nature of the divisions amongst the people who are parties to the agreements, and the bases of these divisions; and third, we identify what the long-term prospects are for the institutions established through such agreements: how far do they mesh with or conflict with wider institutional settings and social changes?

2.1   The basis of the participation

These case studies are of situations in which outside agencies – governments, NGOs, academic departments – have approached local people and encouraged them to participate in the protection of a resource. In some cases this resource is important to the outsiders, but may not necessarily be immediately perceived as significant to those living in the locality (Thomas *et al.*; Panini). In other cases, local people had a definite interest in the resource, but there were no clear institutions to deliver solutions (Chakraborty; Kerkhof), or previous attempts had failed because they were directed from outside and ignored local interests (Kelly).

Two kinds of reasons for cooperation were apparent – first, because local people had the power to disrupt or destroy plans implemented from outside; and second, because the goals could not be achieved without inputs of their knowledge from local people.

The first reason relates indirectly to the issues of conflict. Sometimes these are conflicts between people with different interests in the resource at hand. Kerkhof, for example, notes that farmer–farmer and herder–farmer conflicts have emerged in sub-Saharan Africa. Nomadic herders have routinely been excluded from agreements over the forested areas they use seasonally for pasture, in both the Rural Woodlot Schemes typical of Anglophone Africa and the Gestions de Terroir schemes found in Francophone Africa. Not surprisingly, nomads do not respect the kinds of rules that have been agreed with sedentary populations. But he also reports tensions arising in villages where commercial woodcutters (for firewood) have received privileged access to forests, ignoring the often much large economic returns from other uses of the forest – as grazing, for fruit or for artisanal timber.

These efficiency reasons for participatory agreements raise policing issues: which kinds of groups can ensure that competing stakeholders observe agreements? Local people are regularly and easily able to observe the resource and any infractions of rules, do not have to be paid and can police with consent rather than through force. There are, then, efficiency arguments in favour of involving local people. But, as has been argued for Joint Forest Management in India, this might mean that local people do the work of the Forest Department (Jeffery and Sundar 1999).

On the other hand, some advantages of participation relate to the forms of knowledge held by local people. The notions of indigenous knowledge and its value are highly problematic. Since 1985 or so, debate on the contributions of local knowledge systems to environmental management has been voluminous. Different authors attach different meanings to the terms 'local' and 'indigenous', sometimes reflecting the negative definition of the term as anything that is *not* western or scientific (Gupta 1998: 173; Mathias-Mundy *et al.* 1993). Some (including Sibanda, this volume) have attempted to deal with the problem by using the term 'cosmovision' to mean something more than might be captured by 'world-view' and to refer to ways in which knowledge of the social, natural and spirit worlds are integrated in many cultures. For some writers, programmes of environmental resource management should start from these ways of understanding the environment, and must meet goals defined by local people in their terms (Savyasaachi 1999). Alternatively, 'local' knowledge may be little more than 'knowledge of the locality'. As Sibanda notes, CAMPFIRE, a project that claims to be based on local knowledge, can cope with this latter meaning of 'indigenous knowledge', but is incapable of dealing with issues (raised, for example, by claims about the spiritual value of aspects of the environment) generated by an attempt to go beyond 'knowledge of the local'. The risk of romanticising 'indigenous knowledge' is also clear. Such knowledge is often held by elders, men perhaps, from only one section of the 'community,' who may use their 'ownership' of one version of this knowledge to attempt to reinforce patterns of social control that have been undermined by wider social changes. Social change has also made much of this knowledge theoretical rather than practical, and the notion of an essential core of pure, unsullied people with 'their' knowledge flies in the face of the evidence of the interactions with the outside that have characterised most such communities (Gupta 1998: 172–9).

In practice, many project managers are strikingly uninterested in *any* forms of local knowledge. For those project managers who are aware of the possibility of using local knowledge, what is usually of most interest is the knowledge that people may have of their own localities. This emerges most clearly in the case described by Simpson and Sohani, where information such as whether local trees are a good indicator of the presence of water supplies, where water has been seen in the past and which water courses dry up first in poor rainfall years were of immense help to the design of a successful project. In contrast, in the case described by Lawrence, demonstrating the pragmatic value of local knowledge was useful in convincing certain stakeholders to allow local people a voice. But such knowledge by no means exhausts the forms of local knowledge held locally. Some of these relate to ways of thinking about the social, cultural and symbolic meanings ('cosmovision') that people attach to different aspects of the environment (see the chapters by Sibanda and Panini). In the Pulicat Lake case, Panini shows that project managers had no idea of the complex *padu* system of social organisation; while Sibanda shows that project managers were dismissive of the spiritual aspects of the Tonga relationships with wild animals. Similarly, Biggs, Martin and Matsaert show how scientists in the triticale research programme had for many years been insufficiently aware of local farmers' agronomy and consumption priorities.

2.2 The strength of internal differences within and between user groups

What emerges most strongly from Kabiri's chapter on the Maasai and Niamir-Fuller's chapter on the Karimajong is that environmental conflicts are rarely only about the environment. Environmental deterioration can be turned into rejuvenation only if the people involved have some minimal levels of security of tenure and prospects of a long-term relationship to the resources involved.

Rabindra Nath Chakrabarty and Paul Kerkhof show how, in some less conflictual settings, stable local natural resource management systems may have social institutions that do not necessarily serve the goals of equity, either as between local groups or within them. Stable institutions are often based on existing distributions of power and authority, yet they are capable of managing and protecting aspects of the environment. This finding contrasts with some of the earlier writings in the field. Chambers (1983), for example, argued that successful forest protection committees were those with minimal internal differentiation. Whereas class and gender have been regularly considered, in many settings ethnic differences (by tribe, caste or religion) can also be exacerbated by new institutions. Projects to introduce protection committees usually involve multiple objectives: for example, not just to reduce overfishing but also to ensure biodiversity, enhance livelihoods and reduce the marginalisation of some groups in society. Each goal is likely to be the particular favourite of one or more sets of stakeholders. In such circumstances, those introducing new programmes may have to be innovative in order to avoid having to choose between goals: finding ways to keep fish catches down without increasing the marginality of poor women, for example. While such solutions may be successful in the short run, they are vulnerable to wider processes of social change. Kerkhof notes that old-established groups have moved to take over the benefits achieved by more recent immigrants to an area. On the other hand, if the marginalised do manage to get access to more power, they may choose (at least for a while) to attack the resources they see as the preserve of those who have exploited them in the past.[2]

One of the main groups affected by the change to participatory management may be the government servants who were previously able to exploit their position (whether to extract bribes or to ensure public order in particular ways, for example). Kerkhof notes the opposition of Forest Department officials in francophone West Africa, where the forest services are often organised on militaristic lines; elsewhere (as in the Nepal cases discussed by Chakraborty) the forest service was in a much weaker position to obstruct reforms. In this sense, support for participatory natural resource management schemes may well become part of wider processes designed to 'roll back the state', particularly in Africa where the World Bank has developed most fully its campaigns for 'good governance' as a response to aid failures and economic decline in the 1980s (Bayart 1993; Williams and Young 1994).

2.3 The long-term prospects for the innovations

Biggs, Matsaert and Martin's analysis of the case studies shows the importance of the wider social and economic contexts in understanding whether or not key institutions – like research and extension divisions in the Ministry of Agriculture – are able to play supportive roles in participatory initiatives. In the enthusiasm for the roles of local populations, it is sometimes forgotten that government agencies remain key actors, whether they be Forest, Fishery, Wildlife, Environment or Agriculture Departments, with their associated staff. Apart from the most romantic enthusiasts for 'indigenous knowledge', most commentators foresee collaboration between science-based advisers and local decision-making institutions. Lawrence's paper provides one methodology for moving forward on this front, and points out that some cherished theoretical presuppositions may need to be sacrificed in doing so. The examples in the chapter by Biggs *et al.* show two important ways in which notions of social capital fail to capture the chances of success for such collaborations. As they point out, groups can be strong in the wrong kinds of social capital – the infrastructure of networks, norms and relationships may include the wrong people or stress inappropriate goals. The second problem is that the social capital of diverse groups may be hard to link through formal long-term institutions. Biggs *et al.* stress the need to find non-threatening contexts within which common ground and new coalitions can be forged: one-off events (fairs, conferences, workshops) can provide these opportunities, and there may be no way in which such events can be institutionalised without their losing the characteristics that made them useful in the first place.

3. Wider implications of these case studies

What, then, are the key lessons to be learned from the chapters in this volume? We start from the very important revisions to orthodox institutional economic analyses of the environment, associated with the work of Elinor Ostrom and her associates (Ostrom 1990, 1996, 1999; Ostrom *et al.* 1994). The cases presented in this book fall best into her category of 'coproduction' or 'the process through which inputs used to produce a good or service are contributed by individuals who are not "in" the same organization' (Ostrom 1996: 1073). She makes three further points about coproduction of services:

1. that services usually depend on cooperation amongst several agencies, often across the public private divide;
2. that street-level bureaucrats have autonomy and are not just 'the pawns of a central bureaucratic machine'; and
3. that production of a service is 'difficult without the active participation of those supposedly receiving the service' (ibid.: 1079).

Ostrom goes on to specify the conditions which 'heighten the probability that coproduction is an improvement over regular government production or citizen production alone' (ibid.: 1082). Paraphrasing these conditions gives us situations when participatory solutions are chosen rather than non-participatory ones, as follows:

1. the resources controlled by governments and citizens must be complementary and offer opportunities for synergy;

2. both parties must be legally entitled to take decisions, giving them both some room for manoeuvre;
3. participants need to build credible commitments to one another (e.g. through contractual obligations, based on trust or by enhancing social capital); and
4. incentives must help to encourage inputs from citizens and officials alike.

We will consider each condition in turn.

3.1 The coproductive inputs controlled by each party

In each of the case studies reported in this book, the main under-utilised inputs controlled by local populations are their time, knowledge of the local natural resources in question and skills. The balance of these varies from case to case. Thus Sibanda suggests that most valley Tonga (and certainly the younger ones) have little specific knowledge of the locality and wildlife management, but because of their poverty and marginalisation they have time and some relevant skills. Niamir-Fuller and Kabiri, respectively, note that Karimajong and Maasai have considerable knowledge of cattle and pastoral skills, and labour-power that can be used either to improve or deteriorate the environment. Karimajong and Maasai knowledge can help support very different relationships with wildlife and with general environmental change, depending on wider political and social contexts within which they may be deployed. Lawrence reports on the extent to which short-range variability in agrarian conditions meant that local farmers were best able to see the relevance of experiments; and in one of the case studies discussed by Biggs *et al.*, farmers held the key information about how tools needed to be adjusted to suit local conditions.

On the other side of the equation, it is not always clear what governments can contribute (other than by stopping their unhelpful interventions). In principle, governments (or in some cases NGOs, international or national) have access to superior 'technical' knowledge, and to the capital necessary to introduce some better ways of interacting with the environment. In many of the cases discussed in this book, the solutions are not heavily dependent on either 'science' or capital.

Synergies are most obvious when the different forms of knowledge are complementary. Thus in the case discussed by Simpson and Sohani, finding the sources of water took the combined efforts of very high technology (including the use of Global Positioning Systems and complex soil and water analysis) provided by the University of Waterloo as well as a close attention to local knowledge of watercourses and tree varieties that indicated underground water flows. Biggs *et al.* show how research institutes and extension educators often do have useful knowledge to offer, once they enter into productive relationships with end-users. On the other hand, Panini describes a situation where technical knowledge (of how to create the environments where fish might flourish), while potentially useful, was too narrow, failing to take account of the wider environmental changes introduced by aquaculture and the development of a new port.

The form of any synergy may change as a project or programme develops, depending on who holds what information and who needs the information for what purposes. Kerkhof points out that, as participatory schemes extend to larger and larger areas, it is harder to assess whether or not the ecological results match the claims made in their favour. Scientific methods of forest mensuration are costly, and the skills may be limited to only one or two sources in the country. The need for robust, community-based systems of monitoring is obvious, but hard to establish in ways that command legitimacy from donors and governments (see also the discussion in Harkes 2001). In the cases discussed by Kerkhof, local people probably do have a relatively clear idea of how much resource is available, but once projects move beyond the initial, highly studied experiments, donors and government agencies may have less and less detailed information on what is going on. As Kerkhof points out, forest monitoring (as conventionally understood) is a time-consuming and expensive operation, and the resulting reports are rarely of benefit to the users; SOS Sahel have developed alternative methods of codified monitoring, which are cheaper but still more expensive than existing methods of visual observation and verbal communication of changes. Without such simple mechanisms of feeding back data from the field, the appearance of predictability derived from research carried out in demonstration plots on research farms can be highly misleading, as the paper by Biggs *et al.* shows for triticale.

3.2   Legal entitlements and room for manoeuvre

Where individuals involved in the natural resources are landowners, in all the cases considered they are able to choose to use their land in one way rather than another. Unfortunately, this tends to vitiate the interests of migrant pastoralists. As the chapter by Niamir-Fuller shows, the problems of the Karimajong arise in settings where travel is increasingly regulated and limited, because they are unable to retain a presence throughout the year in the grazing grounds that are essential to the long-term reproduction of their herds and their social organisation. Kerkhof and Thomas provide examples of resources that are small enough for sedentary people to monitor them, but the interests of those from farther afield are much harder to integrate into management plans.

Strong individual rights also impinge on the legally enforceable rights of social units above the level of the household. Kabiri describes a situation where the privatisation of resource ownership has progressed quickly over the past two decades, helping to undermine collective Maasai institutions. In some settings (such as those where wildlife or fish are the resource in question) legal rights may be very unclear. Panini discusses how those who formulated the ecological restoration project for Pulicat Lake saw no need to consult fisherfolk; nor is it clear what was the legal status of the *padu* system by which catches were regulated.

Governments usually see themselves as having the main legal rights, and expect to be able to use such rights to resolve disputes. A first step to the creation of participatory solutions is usually for governments to accept that they can no longer implement their will through regulations imposed from outside – to accept, in fact, that the service in question is coproduced. Genuine transfer of powers – through processes of decentralisation and devolution – are rare, and user-group committees often exist in a legal twilight, as Sibanda shows. Where law and order has broken down (as in the case described by Niamir-Fuller) it may be clear to all that state power is nominal. Where communities live at the edge of governmental reach (in relatively inaccessible areas) state power may be more difficult to implement. But even in such cases (for example, in the Nepali forests described by Chakraborty, and the Sahelian forests discussed by Kerkhof) government agents – typically, forest officers – can choose whether to provide seeds, enforce forest protection measures or turn a blind eye. In several other cases (such as the soil conservation projects in Bolivia and Malawi, discussed by Lawrence and Kelly respectively) governments have not historically either had or attempted to implement decisions.

An early element in several cases of participatory natural resource management is a new law, one that makes it possible for governments to pass over some rights and obligations to user groups or to NGOs. In Nepal, many countries in the Sahel, in the Cameroon and in Zimbabwe, changes in the law were essential but not sufficient grounds for the innovation of cooperative management systems. Thus far, however, we do not have enough evidence to know whether the working of the new laws will in practice allow sufficient independence of action to the institutions they have allowed to come into being: indeed, Thomas suggests that a minimum of 10–15 years may be needed before it is possible to say that participatory institutions are properly established.

3.3 Building credible commitments: trust and social capital

Most of the chapters in this volume describe situations in which individuals and governments are wary of committing themselves and relying on the other. In Uganda, commitments were eroded while Presidents Amin and Obote were in power; however, President Museveni has been trying to rebuild social capital, with some success, as Niamir-Fuller shows. Most of the chapters, however, describe situations with a wide range of institutional arrangements that can be described as showing substantial levels of social capital. They reinforce and maintain significant, dense reciprocal social networks, in which trust can grow and allow user groups to develop norms that ensure that rules are observed and rewards are distributed without generating conflict and causing the institutions to self-destruct. Many of the workshop participants, however, endorsing recent attacks (Harriss and Renzio 1997), were highly critical of the notion of social capital. Such networks could be used to undermine rather than to build: as Biggs *et al.* show, they can point in the wrong directions, forming an almost impenetrable barrier to participatory relationships with those on the outside.

In most cases where projects have been established, donors, NGOs and government departments have attempted to establish relationships of trust from a fairly narrow base. Often, governments have experience of villagers who do not respect boundaries or rules, as Thomas points out for the Kilum–Ijim Forest in Cameroon; similarly, most villagers have experienced governments and officials at various levels who have failed to stand by their promises. Participatory rural appraisal (as used by Lawrence and Kelly) can provide a starting point, as can the 'one-off' events described by Biggs *et al.* The process of negotiating contracts, described by Kerkhof and Chakraborty, can help to overcome some initial wariness.

3.4 Incentives to encourage inputs from citizens and officials

In most cases, the incentives for citizens to be involved in cooperative management measures is their experience of deterioration in their access to the benefits from a particular natural resource. Most natural resource users in developing countries have a low discount rate for future benefits; they are prepared to forgo some current income in the hopes of higher future income streams, even if those future streams are not likely to be very large and are somewhat unpredictable. If the alternative is to abandon current living patterns – to move into town, to turn to seasonal migration or to add further livelihood strategies – people may take little convincing to agree to new rules of access or to help protect or improve a valuable resource.

But users do not all have the same interests, and a move to more participatory management systems often threatens the richer users, who benefit from existing situations. For Ostrom, as for many others (Chambers *et al.* 1989) the existence of differentiated populations, with differential interests in a resource, reduces the likelihood of successful organisation to protect a resource. As Chakraborty's chapter, shows, however, stable participatory institutions may be ones most closely linked to existing political institutions, which are often hierarchically structured. Lawrence's chapter, similarly, suggests that the long-term sustainability of the projects depends on their closer integration with local government. Rich and poor may, therefore, have different interests in a resource: the 'success' of a user group may depend on keeping the rich happy, and such user groups may last a long time and (judged against ecological indicators such as the extent of tree cover or the size of fish catches) be remarkably successful. Other goals espoused by such projects, however, such as empowering marginalised groups, or enhancing the livelihoods of the rural poor, may be left unachieved. Sibanda argues, for example, that although CAMPFIRE may be regarded as a success in general terms, that success is not based on the project claims of using indigenous knowledge or the participation of large numbers of Tonga.

Incentives for governments are not always so clear-cut, often enough the problems of achieving participatory projects are not a result of unwillingness on the part of the resource users, but of the governments who see their interests threatened by such initiatives. It is important here to disaggregate, since different levels of officials may have very different interests. In some cases (as with participatory forest management in India) senior managers may have the incentive of ending conflicts with villagers, returning to profitability and seeing reforestation on lands currently denuded; they may also be enticed by the prospect of large-scale loans and grants, contingent on them including participatory elements in their proposals (Jeffery and Sundar 1999). In India, as in the Sahel (Kerkhof) and Nepal (Chakraborty), village-level foresters may be much less willing to accept the loss of power and income that might result from legitimising village involvement in forest management.

4. Weaknesses of the new institutional economic analyses

These cases, then, suggest some considerable utility to the approaches of the new institutional economics to issues of participation in natural resource management, but they also throw up some problems. We will briefly discuss two linked weaknesses in the approach: a failure to consider the nature and causes of externally-driven changes in the resource; and an inadequate theorisation of the roles of governments.

In the Pulicat Lake example, no matter how much effort had gone into engaging with the fisherfolk on the lake, the project would have been undermined by the introduction of fish-farming (with associated pollution of existing fishing grounds) and the construction of a new port adjacent to the lake. Between them, these developments were certain to transform the fragile physical conditions that produced the right balance of salinity on which the fish thrived. Failure to create or sustain relationships of trust has been disastrous: in July 1999 Pulicat Lake fisherfolk attempting to remove illegal aquaculture farms were fired on by the police. Similarly, in the east African cases described by Niamir-Fuller and Kabiri, the causes of the current situation are a mixture of conflicts over resource use and other conflicts which impact on natural resources. Without dealing with these causes, collective organisations will have little chance of helping to reverse the deterioration.

In both these cases, governments have no interest in empowering resource user groups. In all the relatively successful cases discussed here, governmental interventions were crucial, through the provision of technical support (Simpson and Sohani), legal frameworks (Kerkhof, Chakraborty and Sibanda) or financial inputs to ease the transition to new rules (Thomas *et al.*). Such supportive interventions can emerge only when the interests of powerful stakeholders are not threatened. These need empirical investigation in each setting. As we have already pointed out, governmental structures must be disaggregated, since the interests and attitudes of different levels of staff may be very different (Vira 1999; Jeffery *et al.* 2001); at a wider level, local-level analyses must be located in a local, national and global political economy (see, for example, Biggs and Matsaert 1999; Jeffery and Sundar 1999).

5. Conclusion

The case studies reported in this book do not provide just an overview of some recent projects in the field of participatory natural resource management, they also advance the discussion of the kinds of steps needed to raise the quality of such projects and to enhance the chances that they will meet the multiple goals that they nearly all espouse. The dominant paradigm, drawn from institutional economics, can be seen to provide a very useful framework to analyse relatively special cases: those where the roles played by external social actors (especially governments) are minimal and/or benign; where the physical environment is stable or supportive; and where the causes of environmental deterioration can be addressed by the user group and do not require wider changes to take place.

These insights draw attention to the fact that the claims made for participatory projects should be limited: participatory projects play an important role, but are inevitably only a part of any long-term solutions to problems of natural resource deterioration in developing countries. Analysis of participatory projects must go beyond the idea that they are the 'silver bullet' that will resolve every difficulty, and that in their absence the environment will continue to deteriorate. More modest claims – even for apparently 'successful' projects – will ensure that the issues of social inequality, which are usually the root causes of environmental deterioration, will be more clearly addressed in future.

Notes

1. We gratefully acknowledge the financial support of the Economic and Social Research Council's Global Environmental Change Initiative, and the moral support of Alistair Scott in particular. The workshop was organised by the two editors and held in Mansfield College, Oxford in April 1998.
2. See, for example, the activities of the People's War Group in Andhra Pradesh, eastern India, who for several years excluded not only the Forest Department from their areas of control but also called for clearing of forest areas for agriculture, and attacked representatives of NGOs who wished to encourage local forest protection committees (Jeffery and Sundar 1999; Yadama and DeWeese Boyd 2001).

Part I

Where Local Conflicts over Resource Use Make Participation Unlikely

2
Conflict Management and Mobility among Pastoralists in Karamoja, Uganda

Maryam Niamir-Fuller

1. Introduction

Early anthropologists have documented a history of shifting alliances and tribal warfare in Karamoja. Uganda's political troubles in the 1970s and 1980s aggravated a fluid situation of raids, conflicts and revenge killings. In the 1990s, under the regime of President Museveni, experimental, innovative conflict-resolution mechanisms have been tried out. The internal and external stresses causing the conflicts are as strong as ever, and solutions need to be found quickly to prevent the situation from deteriorating further.

In this chapter I detail the dynamics of conflicts over land, mobility and people's access rights to the natural resources. I briefly trace the trends of increasing poverty, proliferation of guns, and erosion and fragmentation of the authority of traditional elders. In considering the sources of conflicts in Karamoja, and traditional and emerging conflict management mechanisms, I discuss whether these mechanisms provide design principles on how to define a 'community' for natural resource management and how social capital can be reinforced.

2. Background to the Karimajong

Karamoja is located in the northeastern corner of Uganda, bordering Sudan and Kenya. The long-term average annual rainfall (1939–95) is 620 mm, but it can vary between 420 mm and 1,260 mm, with a coefficient of variation of annual rainfall greater than 30 per cent. The distribution of rainfall in time and in space is the main determinant of the ecological potential of this land. Three major droughts preceded the twentieth century: the first recorded in traditional folklore occurred between 1706 and 1733, the second about 1800, and the third from 1876 to 1900 (Pazzaglia 1982), during which the estimated loss of cattle was between 70 per cent and 90 per cent. Between 1924 and 1952, almost every third year resulted in crop failure due to droughts (Dyson-Hudson 1966: 76–7). The 1960s were relatively good, but since the 1970s, crop failure has returned to its three-year cycle. In addition, rainfall is not always distributed adequately during the wet seasons, leading to false starts and flooded harvests. The uncertainty of the rainfall is the primary reason why the Karimajong do not depend solely on cultivation for their livelihood, and why they consider the land better suited to livestock production.

According to the 'non-equilibrium' theory of ecological systems (Ellis and Swift 1988) arid ecosystems never achieve equilibrium because of the high degree of variability. The ecosystem is constantly changing from one level or state to another. Thus, it is difficult to predict vegetation responses to stresses such as overgrazing, fire, drought. Furthermore, sustainable use of the resources must be flexible enough to follow the constant climatic changes and the 'patchiness' or heterogeneity of the ecosystem. The Karimajong have learnt that mobility of livestock, on both a daily and seasonal basis, can use the patches to their fullest. Estimates of carrying capacity average out high and low quality micro-niches, and result in figures below the actual potential. Tracking,[1] dispersion and mobility are variables with which the pastoralist can viably stock the rangeland at higher rates than that suggested by an average carrying capacity figure.

The 'Karimajong cluster' of people today are defined as those pastoralists and agropastoralists that occupy the Province of Karamoja in eastern Uganda, whether they be Nilotes, Kalenjins or Bushmen. They share common social and economic features through years of assimilation: an age-set system, the belief in a deity, and the pre-eminent social and economic value of cattle. They continue to perceive themselves as separate tribes, however, each claiming its own land and backed by its own armed forces. The production system of the Karimajong varies with the ecology, but in general they are agropastoralists, raising both livestock and crops. The Karimajong have one feature in common: the feeding requirements of the livestock pull the herd owner to peripheral rangelands, but the security requirements of the family pull it to the centre where the bulk of the settlements and farms are. Herd owners have reconciled these conflicting needs by splitting the herd into a main herd and a milk herd (Dyson-Hudson 1966). The head of household has to ensure that the settlements have enough milking cows to feed the settled population, but not to leave too many to attract raids.

The Karimajong ideal is to possess a large enough herd to support an extended family of three generations, who in turn provide the labour for the herd and crops. They do not consider wealth in a speculative manner (as do capitalistic entrepreneurs) but as a form of economic security over the long term. The frequent droughts create a 'boom and bust' situation, forcing the owner to raise a large enough herd so that a core group of surviving breeders can reconstitute it after a drought. The livestock owned today by the Karimajong has been greatly exaggerated. Total livestock population in the early 1990s had yet to reach the pre-1959 level. Furthermore, the average herd size per household has decreased because of human population increase. Between 1900 and 1950 this average is estimated to have fallen from 100 to 50 heads per household (Rattray and Byrne 1963). Today, the ratio is no more than 28 (assuming ten people per households). By contrast, the Karimajong ideal of a self-sufficient, viable household economy is six cattle per capita (Baker 1974), or at least 60 heads per household. The average household at the moment is currently below the level considered adequate for long-term economic viability.

Data on income distribution are scarce, but indications are of relatively minor income gaps between households in Karamoja. A small elite has emerged, consisting mainly of educated Karimajong who have left the livestock industry. The relative homogeneity and lack of distinct classes may result from the historical patterns of raiding, followed in recent years by ad hoc banditry, as described below.

The Karimajong have never relied solely on the resources of Karamoja proper, but have always taken advantage of natural resources outside. Examples are the Acholi and Teso lands to the west, and Turkana land in neighbouring Kenya to the east. Transhumance (seasonal movements) to these areas usually followed negotiation and agreements with the respective tribes. In recent years, this mobility has been curtailed through administrative decisions and political insecurity. In addition, grazing lands are increasingly coming under cultivation, because of the relative security (in terms of raids) of crops as opposed to livestock. Land degradation and resource scarcity have been a direct result of these trends, as more and more people and livestock congregate on less and less land.

2.1   Socio-economic and political significance of raiding

Raiding is a traditional activity among all plains Nilotes. In the past, its primary purpose was to regenerate the number of animals in the herd as a response to drought. Therefore, it was sanctioned by society and given spiritual and material support by all (Ocan 1992). The

Karimajong were not criminals in the sense of robbery or rape; they were warriors on a noble mission with the blessing of their tribal spirits (Rattray and Byrne 1963). In the past, carrying out a raid without the full backing of the elders and the divine was unthinkable. Traditional leaders of successful raids gained much prestige from their prowess, and apart from being the choice catch for unmarried women could become potential organisers of future raids. Some notable raids have been recorded in Karimajong oral history, such as that between the Jie and Bokora at Nakoret Amoni, near Panyangora (Lamphear 1976: 205–8).

Although guns have been used since the 1880s, violence and banditry did not explode until the 1970s (under President Amin) and the 1980s (under President Obote). As an officer in the colonial Kings African Rifles, Amin pursued the Karimajong and Turkana ruthlessly. As President in 1973, he ordered the mass killing of hundreds of people in southern Karamoja (Ocan 1992: 16). In 1979 after the collapse of the Amin regime, which coincided with a major drought, the Karimajong broke into military armouries and equipped themselves. The same occurred after Okello's regime in 1986. A new pattern of heavily armed raiding began to emerge.

Ethnic conflicts are now sharper than ever because of privately motivated raids (Belshaw *et al.* 1996). Not all raids are preceded by ritual and community consensus. Many raids are organised on an ad hoc basis by individual warriors, mostly for material gain. No longer is only livestock stolen as it once was: now even household goods are taken. Poverty is partly to blame, as is increasing 'individualisation' of the society.[2] Banditry on the roads, aimed mainly at passenger vehicles and buses, was a major detriment to the local economy and led to suspension of most development aid in the late 1980s and early 1990s.

One defence against raiding is to vacate remote areas and settle close to population concentrations. Maps of the 1960s show that the southern part of Jieland, bordering on Bokora, was fully settled, but these lands have now been abandoned for Kotido Town and the area to its north, leading to greater land pressure and degradation. Written and oral reports agree that the Jie would use the eastern zones (bordering Turkana) as wet season pastures (Gulliver 1970). By the 1990s, with renewed conflicts with the Turkana, these areas were also abandoned. The Dodoth, Ik and Nyangia have adapted differently to raids. They prefer to disperse into the crevices and valleys of the mountains to the north, venturing into the plains only to cultivate rapid, short-season crops. These land use changes have resulted in land degradation in Karamoja. Overgrazing can be seen in the mountains of Dodoth; and in the case of the Jie, Bokora and Matheniko, over-concentration of settlements has created overgrazing and deforestation around them.

3. Local institutions

Apart from the sole attempt by the Jie to forge a centralised authority in the 1880s, the Karimajong remain essentially fragmented tribes and sub-tribes. Traditional political authority is based on the age-set system (Dyson-Hudson 1966). All Karimajong adult males progress through a series of age-sets and generation sets in their lifetime. Five age-sets comprise a generation set, of which there are two: elders and juniors. The elders, responsible for the community's welfare, are authorised to direct its activities. The juniors are subordinate and are the instruments of policy, such as conducting raids.

Men take all the decisions regarding political issues, war, alliances and cattle movements; women are the decision-makers over daily life. All family, children and village activities are women's responsibilities. Recent efforts to introduce mobile education adapted to pastoral needs among the Karimajong was only able to proceed once the women, in a major gathering, agreed to allow their children to attend and to adopt modern education. They were then assisted by the elders in a ceremony called 'Unearthing the Pen'.[3]

In addition to the age-set, the Karimajong recognise a clan system based on heredity. A clan is headed not by one, but by as many elders accepted into that generation from that clan. Clan members are scattered among different locations and tribes. The same clan name is to be found among the Jie and Toposa, as among Matheniko and Bokora (Pazzaglia 1982). Clans in turn are divided into sections called *ekitela*, which are territorially based and associated with the settled agricultural areas. The *ekitela* are institutions primarily for organising religious and social events, but also for taking production decisions. In a conflict among persons of different clans or sections, one usually sides with the people of one's own section rather than those of the clan, because territory and land use are better determinants of everyday cooperation and loyalty (Novelli 1988).

Each section has many 'kraal leaders'. These are younger members of the elder generation, or older members of the junior age-set. Kraal leaders are in charge of the main herds of the section and take responsibility for all decisions on mobility, camp location, scouting, watering, labour allocation, dairy cow distribution, etc. Several herders work under the kraal leader, depending on the size of the herd. The kraal leader is selected by consensus, on the basis of his prowess, intelligence and social position. The *aruanit* are highly respected kraal leaders who have a special position in all public debates. The Jie, for example, have 21 *aruanits*, and the Dodoth 30. Each year kraal leaders will decide which *aruanit* to join for the dry season movements, or whether to strike out alone. Although movements and yearly allegiances of the kraal leaders are fluid and opportunistic, each *aruanit* has his geographically specific movement pattern; the starting point is always the settlement area of his section, going through specific water points towards a specific dry season grazing area. The *aruanit* will not waver from this pattern, except to change the speed of movement or to decide how far into Acholi and Teso land to venture.

Beginning in 1916, the British Administration imposed a hierarchical system of chiefs based on the Buganda model from southern Uganda. The chiefs and sub-chiefs were given various administrative responsibilities, among them to enforce the dress code, compulsory education and to collect taxes. Dyson-Hudson (1966) remarks that imposition of such a new form of organisation met with failure since it never obtained the approval of the elders. Until the Second World War, the Colonial Administration did not consider that Karamoja could contribute to the economy of Uganda, and its policy was based on containment, pacification and sedentarisation of pastoralists. By 1945 'development' of pastoralists replaced the old view, and for the first time a Veterinary Officer was stationed in Karamoja (Rattray and Byrne 1963). The early postcolonial period saw the continuation of Colonial Administrative structures and policies. But the political events of the 1970s and 1980s led to the breakdown of law and order. Government administrative organs barely functioned, rural development was forgotten and militarism took over.

Local government today is based on a hierarchical system of districts (of which there are two in Karamoja: Moroto and Kotido), divided into counties, sub-counties, and parishes. A local government representative is present at all district and some county levels. The government of President Museveni has initiated a system of local 'Revolutionary Councils' (RC) at each administrative level, which are supposed to elect members from the local population. In practice, Council members are usually young, educated and/or urban-folk (traders, teachers), who have relatively little contact with the traditional leadership.

As banditry became more prevalent than community-sanctioned sorties, elders saw their powers gradually ignored. Other factors related to the imposition of a government administration and judiciary, and nationalisation of land. The authority of the elders was eroded, and they lost the ability to control the activities of the young warriors. Gradually a few of the kraal leaders, gaining more successes than others through banditry and raids, have become the so-called 'warlords' of today. These shadowy figures are behind most of the organised banditry in Karamoja.

The national government structure after independence was never able to come to grips with the situation in Karamoja. Apart from a well-armed military presence, the administration was weak, with few, poorly motivated personnel. With the coming to power of the National Resistance Movement, the 'problem' of Karamoja became a national priority. After 1986 the National Resistance Army of President Museveni gradually restored order in the land and the internal supply of guns diminished. Today, guns still filter through from southern Sudan, and after the overthrow of Siade Barre in 1990, from Somalia via Kenya and Ethiopia. The collapse of Mengistu's regime in 1991 saw additional guns and ammunition in the market.

Early development efforts by the current government focused on the equitable distribution of dry season water points ('valley tanks'). A Ministry of State for Karamoja was created, with an implementation mechanism called the Karamoja Development Agency (KDA). This institution was to coordinate all development activity in the region. These early efforts have not had much success. The causes of failure have been identified as insecurity, little or no community participation, public/open access to the water source, and poor assessment and analysis of the basic problems facing the Karimajong (KPIU 1996).

Beginning in 1994, with the appointment of a new Minister of State for Karamoja, a Special Presidential Adviser for Security Issues, and a Brigade Commander, all of whom are Karimajong, a greater convergence between the Karimajong reality and government policy is being achieved. Another emerging institution is democratic representation in parliament. The 1996 parliamentary elections brought to Kampala four very active and young Karimajong, whose expressed mission is peaceful resolution of conflicts, and rural development in Karamoja.

Leadership among the Karimajong has now diversified and political power has been dispersed. At 'higher' levels of customary organisation, the age-set system has lost its inner political strength, although the outward appearance is still strong and religious and cultural ceremonies are still practised. The 'lower' levels of customary organisation, at the camp and kraal levels, are on the contrary quite strong and determine local political decisions as well as raiding patterns. Thus traditional leadership has disintegrated among elders, *aruanits*, kraal leaders and warlords. At the same time, modern leadership is trying to establish a niche of its own (RC leaders, local administration, parliamentarians, intellectuals). There are too many leaders now and authority has fragmented. Most leaders wish to see an end to the conflicts, but they do not agree on how it should be done. The strength of the moment is that such a fluid situation allows experimentation with different solutions.

4. Sources of conflict

By far the greatest causes of conflicts in the past were land ownership and land use issues. These continue to occur, but recently, conflicts in the form of banditry for individual gain have gained ground.

4.1   Land tenure and land use patterns

Customary land tenure in Karamoja is organised around membership in tribes. Each tribe has its own clearly defined 'home territory' where the settlements are located. The home territory also includes all wet season and some of the early dry season grazing areas. The boundaries of the home territory are more or less fixed. Successful raids can sometimes expand the boundary, but the new territory is used for grazing, not for settlement or cultivation. Open, natural sources of water are free to all tribal members. Outsiders must seek permission from section elders to use the water and the land around it. Improved water sources (hand-dug-wells and *ngaperon* or hand-dug small catchment basins) are the property of the individual or section that invested in it.

The early policy of the Colonial Administration was to stabilise pastoralists within strict boundaries. They defined these boundaries according to where they found the people at one given time (Gulliver 1970: 9). As a result, the pastoralists were required to restrict themselves to only one of their seasonal migration areas. The boundaries were patrolled by the Colonial Administration with the zest and vigilance normally reserved for international borders, because Karamoja was still considered to be an 'occupied territory' (Mamdani *et al.* 1992: 29). These efforts were met with resistance and defiance from the various groups.

One of the main sources of conflict in Karamoja has been the perturbation of patterns of land ownership and access rights. Karamoja's border with Kenya was first delineated in 1940 during the colonial era. In so doing, the Karimajong formally lost about 15 per cent of their southern grazing lands to the Pokot of Kenya (Dyson-Hudson 1966: 783), thus sparking off a major conflict that still lasts to this day. The Karimajong also lost their *de facto* rights to use the fertile dry season grazing land in Teso (Rattray and Byrne 1963: 8; Cisterino 1979: 79), and only do so these days when they have to, and in stealth. Another major factor was the cordon sanitaire, established on the border with Sudan in 1919 to contain Sudanese rebels. Settlements near the border were relocated further south, creating more resource pressure. Wild herbivores spread into these abandoned areas, bringing with them tsetse flies into Karamoja for the first time. By 1948 the fly had occupied all the grazing grounds up to Labwor, in spite of a 32 km bush and tree clearing around Kotido to halt its spread.[4] This pushed the Jie and Dodoth out of many of their dry season pastures into Labwor, Mening, Ik and Bokora lands, causing additional stress, resource competition and conflicts.

Another factor affecting land use patterns between the tribes was the construction by the Colonial Administration of at least 108 dams and water-retaining structures, mostly in the central plains and the western flatlands. The rationale was to increase the land area available for dry season grazing within Karamoja and to avoid the need to use Acholi and Teso lands. The structures were put in place and seen as public goods, therefore open access to all. This could have resulted in a 'tragedy of the commons', but not everyone changed their grazing movements, because of inadequate forage around the water points. The postcolonial government was not able to maintain the structures because of financial constraints, and more than 90 per cent fell into disrepair. A further factor that reduced the rangelands of the Karimajong was the establishment of forest and hunting reserves, and National Parks. By 1958, 25 per cent of Karamoja was gazetted by the Colonial Administration (Mamdani *et al.* 1992: 47).

All in all, 40 per cent of the original lands of the Karimajong (not counting rights to Teso) were taken away during the colonial period. The Karimajong are now restricted to only a small portion of the remainder, because of tsetse flies, to avoid the Turkana and Pokot, to avoid conflicts with the Teso, and because of increasing insecurity among the Karimajong. The mobility of the Karimajong is no longer determined by non-equilibrium ecological forces and political alliances, but also by increasing insecurity and conflicts between tribes. This change has resulted in land degradation, for the moment only around settlement areas, and in lower productivity of animals, because of the inability to properly 'track' feed resources.

Today's administrative boundaries in general coincide with the home territories of the major tribes, but they do not include all their annual grazing areas, which sometimes cover several districts or provinces. In addition, minority tribes (such as Mening, Nyangia, Ik) are subsumed under the main tribes, potentially causing conflicts. The Labwor District (part of which is traditional Jie land) has more land than the Labwor can use today because they no longer herd livestock as before. Only in recent times have administrative boundaries begun to make an impact, and only because of the adoption of democratic rule in Uganda. Now, parliamentarians are elected from each District and County, and gradually the majority of Karimajong are beginning to understand the political significance of such boundaries, if not their significance for land utilisation.

The 1995 Constitution of Uganda attempted, under Article 237, to clarify land tenure in Uganda, setting out the processes of private ownership and acquisition of land, and clearly empowering the local government to control those lands that are for the common good of the citizen. The constitution recognises four land tenure systems, all related to private ownership of land. It does not explicitly recognise communal ownership of land but it can be assumed under 'customary tenure'. It is silent on the issue of mobility of animals on common land, and the need for negotiable 'inclusive', not exclusive, rights to grazing resources.[5] Although formal laws are indeed 'a thing far away' for the Karimajong, they nevertheless can offer the backdrop for developing new forms of common property regimes. The land law of Uganda is far from adequate in meeting this need.

The Karimajong gains access to grazing land through negotiation on a yearly basis. An alliance between two tribes, once it is obtained, is a transient, but *de facto* recognition of the right of tribal members to use each other's land without specific permission. Alliances, however, can be fragile, and a herder exercises his rights with caution. The 'rural radio' (word of mouth) is an effective means of information dissemination in Karamoja, through which the herder keeps abreast of all raids and alliances. The pattern of raiding is not one-on-one revenge, but opportunistic raids against any neighbour. Whoever has plenty of animals should expect to be raided by anyone. Raids in the past, however, respected the accords and alliances established by elders. One did not raid the livestock of a 'friend', unless there were signs that the alliance was breaking down, or unless the elders made the collective choice of breaking the alliance. With the breakdown of the elders' powers, such alliances are no longer a major determinant of the pattern of raiding by the independent warlords.

5. Conflict management: prevention and resolution

5.1 Traditional mechanisms for conflict management

The Karimajong, like other pastoral communities, have developed many formal and informal mechanisms to prevent and resolve conflicts. Not only are these mechanisms an essential part of the social capital of the society, but also the entire 'package' of social capital requires their support. These mechanisms were never static: because of their inherent flexibility, they were always adapting to changing situations, even before the advent of colonialism. As these mechanisms are common among all Karimajong sub-tribes, they provide a common set of tools to resolve and prevent inter-tribal conflicts. These mechanisms survive in a weak form today, not so much because of the impact of colonialism (in fact the policy of pacification may in large part have strengthened them), but because of the impact of the postcolonial actions described above.

Informal sanctions, reciprocity, spontaneous adaptations and negotiated alliances are all mechanisms by which the Karimajong seek to prevent conflicts. Informal sanctions by the community, through the threat of social ostracism, ridicule, satire and other overt or covert demonstrations, are flexible and powerful means by which the community prevents the formation and/or escalation of conflicts and ensures communal discipline. Even though formal rules have broken down in Karamoja today, informal sanctions continue to function, but they too will be ignored if the basic fabric of the community and its social capital is destroyed.

Another mechanism for preventing conflicts is the social convention surrounding reciprocity of favours. Reciprocity is the backbone of the community spirit among the Karimajong. Giving and receiving permeate their life. Receiving entails the creation of a new bond requiring gratitude or a reciprocal exchange, leading to a web of supports. Friends are for giving and receiving, while enemies and all others are for exploiting (Novelli 1988).[6]

Conflicts are also prevented by adapting to the stress leading to the conflict. For example, the people of Labwor, after a series of raids and incursions, completely gave up livestock by the 1980s. They could do so since the Labwor Mountains and valley ecosystems were of higher agricultural potential and the people could subsist almost entirely on crops. Only recently have the more affluent people of Labwor started to reinvest their surplus into cattle, but even they have not revived their tradition of herding, and would rather entrust the care of their cattle to the Jie.

The relationship between the Jie and Labwor has swung from one of overt conflict to one of negotiated alliances and trust.

Another form of spontaneous adaptation has been by the Ik, who have concentrated more on hunting-gathering and cropping, although they continue to raise a few livestock. Caught between the Dodoth and the Turkana, herding has never been easy for them, but they have learnt to play off one side against the other. Thus, most conflicts are resolved on an ad hoc basis, through skill, political expertise and luck. This does not, however, protect them from occasional severe raids and attacks, misunderstandings and reprisals (Ayoo 1995).

The Karimajong also seek security from conflicts through alliances with the greatest number of groups possible, via the exchange of gifts and reciprocated favours. Alliances, operating at the level of individual herders, women and even children, as well as at the level of sections and tribes, guarantee survival (Novelli 1988). Such alliances and enmities are fluid, with necessity and opportunity dictating their composition at any point in time. At present the Jie are friendly with the Matheniko, Labwor and the Ik. They continue to raid the Dodoth and the Bokora and the Turkana. However, the Turkana are friendly with the Matheniko but not with the Jie. Potential powerbrokers and mediators could be found within this fluid system. For example, the Matheniko could potentially mediate between the Jie and Turkana, but this role has not yet been exploited.

Alliances are established through the offices of the tribal elders. Little is known about this traditional process, how often tribal elders meet and at what level or scale they are established. More analysis and description of this important process could help identify elements for the design of an adaptable process for arriving at agreements and alliances.

Traditionally elders exercise their judicial power through neighbourhood, sectional or tribal assemblies, called *akiriket*, and in various specialised tribunals. The *akiriket* is a formal, prescribed event held periodically. It has jurisdiction over such issues as raids, sacrifices for rain, returning cattle after raids, and possible new grazing areas. An *akiriket* also provides an opportunity to restore peace between individuals. The *akiriket* is democratic in that even the uninitiated are allowed to attend, but elders always have the last word. The main speakers are never the same, but are either chosen or may volunteer. It is held periodically on an ad hoc basis whenever the need arises. Thus, as an institution it is formal, permanent, and has set rules, but its process is very flexible.

Tribunals are held when needed to resolve conflicts between individuals. There are separate tribunals for men and women. Women's courts are presided by older women. Conflicts covered by the men's court can range from social ones to conflicts over land and resources (for example, planting in another's fields, using another's water without permission). Clan tribunals, under the authority of clan elders, are also held to pass judgement over family problems and administrative matters (Pazzaglia 1982).

The objectives of tribunals in Africa are to reconcile disputants and to maintain peace, rather than to punish the wrongdoer (Rugege 1995). Achieving consensus publicly (accompanied by social censure and ostracism) is more important than providing a 'lesson' through punishment. The 'winner takes all' judgements favoured by adversarial systems of law, are generally avoided in favour of a 'give-a-little, take-a-little' principle (Cousins 1996). Rights of appeal exist, but are seldom used.

In this system, a guilty person is handed over to the elders, judged and punished in proportion to the crime committed. The Colonial Administration imposed its own, British-based judicial structure on this customary system. When they did catch raiders, for example, they applied their cumbersome procedures with its intricate subtleties, instead of the straightforward and swift Karimajong practices (Novelli 1988). The Karimajong soon lost patience and respect for the modern judicial practices. The British system continues today in Uganda, and the Karimajong continue to distrust it.

These Karimajong mechanisms are all flexible in use. Some are informal (social sanctions, reciprocity, for example) and others are formal (tribunals, for example). Among those that are formal, some are transient and are expected to change with time (adaptations to stress, alliances) and others could be constitutional (long-term alliances). As long as there are alliances, conflicts between the 'friends' are minimal, and the breakdown or absence of traditional regulations for resource allocation and political power is neither noticeable nor destructive. But the absence of traditional regulations makes conflicts between 'enemies' appear sharper than ever before.

5.2   Recent mechanisms for conflict management

In the 1980s and 1990s, the trends towards decentralisation, deregulation and popular participation have offered important insights into the analysis of conflicts over natural resources all over the world. A move away from formal judicial processes has led to local level, more participatory processes, such as conciliation and mediation, or what is known as 'Alternative Dispute Resolution' (ADR) processes (Cousins 1996). In Karamoja too, there is a movement towards mediation for peace talks and consensus-building.

Postcolonial governments were faced with serious conflicts along the international borders and tribal boundaries. By the early 1980s these conflicts were so severe that at least five meetings between administrators and people were held in all districts neighbouring Kenya, including Acholi and Turkana. None achieved a lasting solution (Ocan 1992), primarily because the traditional leadership was not explicitly and formally involved.

The current government policy is to prevent conflicts between Karimajong, Turkana and Acholi by continuing to push for the stabilisation of each tribe inside its own provincial boundary. Estimates of the carrying capacity of the dry season pastures in Karamoja – albeit 'guesstimates' useful only for planning purposes – indicate plenty of 'space' within Karamoja, in a normal rainfall year, to accommodate all the Karimajong herds, if and only if the land is used without undue restrictions on internal mobility (that is, opportunistically). Insecurity makes this impossible at the moment. Reducing insecurity will in the long run open up new pastures and allow the dispersal of grazing pressure for a more optimal use of land. However, the Karimajong, as well as the Turkana, will continue to need to use their neighbours' lands in the event of drought or bad years. Therefore, the policy to stabilise each tribe within a well-defined land area will work in good or normal years, but will create hardships and conflicts in bad years. Formal mechanisms are necessary to allow negotiated access in bad years, and thus avoid livestock loss and conflicts. Land use planning and regional development, incorporating principles of inclusivity, negotiated access, flexible and opportunistic resource use, and equitable distribution of resources, would contribute considerably to the prevention of land conflicts.

Recently, Karimajong intellectuals have embarked on a major campaign to promote education among their people. They see education as necessary not only for rural development, but also for increasing security and preventing conflicts. The Karimajong, they believe, should be assisted to find alternative ways of securing livestock and livelihoods, and not to rely on raids and banditry. Alternative livelihood systems need to be found not only for the disenchanted youth and bandits, but also for the herder and his family, in order to secure his livelihood and prevent conflicts.

The Karamoja Task Force, established within the Ministry for Karamoja in 1995, has been instrumental in lobbying and creating the necessary environment for peace talks, brokered negotiations and agreements, and fund-raising through NGOs and donors for Karamoja. It works closely with the Karimajong parliamentarians. It is a transient institution and is expected to disband once its usefulness runs out.

At the moment, however, it can step outside government politics and act as a neutral initiator of equitable development.

A government initiative in 1994/95 encouraged independent warriors to become vigilantes under the command of the more respected kraal leaders. The vigilantes keep their guns and receive a monthly stipend and a shirt symbolising their status. The programme has apparently had considerable success in attracting the warriors, and in reducing the incidence of banditry and, to a lesser extent, communal raids. Locally, people know who committed banditry and raids, and the vigilantes, Karimajong themselves, can easily catch the perpetrators and their loot. For all its perceived merits, the vigilante system may have the unintended effect of legitimising the wealth gains and political power of the stronger groups who have more vigilantes and guns. This may lead to a weakening of the bargaining position of the smaller groups in negotiated peace agreements in tribal talks.

When perpetrators of raids are caught by the vigilantes, they are turned over to the local authorities, and not to the traditional elders system, because of the direct involvement of the government in this process, and because tribunals and *akiriket* have been unable to cover new situations that have arisen, such as banditry. However, moves are underway to strengthen and rebuild the power base of the elders to take over this function. Elders' Councils in each tribe are now being convened on an ad hoc basis, usually by parliamentarians and district commissioners, to resolve inter-tribal problems during 'peace meetings'. These Councils should be modelled after the *akiriket* and tribunals. If successful, this institution-building process can strengthen the power the elders have over all internal problems (rangeland use, grazing rights, cropland, mineral rights, and so on).

The process of conflict prevention and resolution is just as important as the goal. Agreeing to go to a negotiating table itself is an important concession and commitment. The process should be kept as flexible as possible to allow new forms of institutions and mechanisms to emerge. Conflict management theory provides alternative processes and procedures for settling and resolving conflicts, which should be tailored to specific situations. Some 'diagnosis and design' questions adapted from Cousins (1996) may be useful in Karamoja:

- What is the nature of the situation (type of conflict, causes, phases, nature of stakeholders and their political power, etc.)?
- How can the pre-negotiation process be managed (collaborative or participatory planning, empowerment strategies, equalisation of power of parties, information gathering and communication, etc.)?

- What processes and procedures should be selected (culturally appropriate, gender and equality sensitive, integration of customary processes, role of mediators, sequence of processes, etc.)?
- What institutions and mechanisms need to be created or strengthened to sustain the process and the resolutions (transient or constitutional institutions, strengthening of customary structures, capacity-building, etc.)?

One conclusion of the recent revival of peace talks in Karamoja is that inequality limits the usefulness of negotiation. District officials and parliamentarians aware of this fact invite all stakeholders, assisting minorities and those in remote areas who do not have as much political power as the rest. This observation has also been made elsewhere (Bradbury *et al.* 1995; Ross 1995). Attempts to equalise power can be done through various empowerment and capacity-building measures (Cousins 1996):

- modifying the procedures used to manage or resolve the conflict, to ensure equal access by all (for example, meeting in settings where the weaker party feels comfortable);
- legal advocacy and political action to change the legal framework of rights to resources (for example, land tenure laws, political representation in councils and parliaments, dissemination of legal information, etc.);
- mobilising and organising strategies to press claims and defend rights (for example, organisation of new associations, participatory decision-making, workshops and public debates, etc.).

An attempt to equalise power is an important but extremely sensitive process, which may backfire if the stronger parties raise objections or try to retain their advantage.

6. Conclusion: the strength of social capital and the choice of institutions

The most effective approach to conflicts in pastoral areas is to maintain flexibility and to recognise the complexity of the problems. Changes in mobility patterns, constraints imposed by a reduction in territory, decreasing herd sizes and increasing population have all resulted in lowering the standard of living of the average Karimajong. Herders are no longer able to exploit the variable natural resources properly, leading to decreasing livestock productivity. Increasing poverty, proliferation of

guns and erosion of the power of elders have led to an increase in insecurity, banditry and other conflicts. Resolution of the conflicts in Karamoja must benefit from both preventive and curative mechanisms.
The social capital of the Karimajong can be expressed in five ways:

1. cultural and religious mores and values that touch both formal and informal interactions;
2. indigenous technical knowledge (ITK) covering the dynamics of natural resources and people's ability to harness them;
3. duties and responsibilities, depending on one's position in the socio-cultural hierarchy;
4. sense of social discipline back by a political authority;
5. conflict management mechanisms.

Of these, the sense of social discipline has been the first to weaken, as the political (but not moral) authority of elders has weakened. Duties and responsibilities that touch upon production, economic survival and religious interactions are still quite strong, but not the duties one traditionally owed to political leaders. Consequently, only those conflict management mechanisms that depend on political authority (that is, formal conflict resolution institutions) have also weakened. Other elements of the social capital, particularly ITK and cultural mores, are still very much intact. In other words, the elements of the social capital that reflect socio-cultural and economic needs have not weakened, while those that depend on the customary political structure have done so. This dichotomy and the fact that most elements of the social capital are still viable despite conflicts and banditry, can be explained by the fact that political authority has not entirely disappeared, but only fragmented.

Karamoja provides an interesting study as one of the few cases where the internal ingredients exist for spontaneous experimentation with conflict resolution alternatives. The conflicts have approached a 'crisis' phase, thus prompting all stakeholders to take responsibility for their commitments. The national political climate is favourable to democratic and participatory processes. A core group of determined, committed and more or less neutral mediators are present to push the process along. A group of elders can analyse the extent of the loss of their authority, and therefore make remedial decisions, and most people perceive the need for change.

An interesting dimension to development aid in Karamoja can be seen from the people's own eyes. The early generation of development workers, following on the heels of colonial administrators and missionaries, came with a top-down paternalistic approach, hardly understanding the Karimajong mindset. Development workers were seen as strangers, and therefore neither friends nor enemies, but all others, who should be looked upon with indifference and exploited. What development aid lacked in the past was the ability to transform itself from all others to friends; such a transition would have assured human dignity to the development activities – the possibility of 'reciprocating' assistance (Novelli 1988). Only through a clear understanding of the Karimajong culture and world-view can true participatory, self-sustaining development be carried out.

Conflict management mechanisms are suited to different levels or scales of conflict. For example, the *akiriket* and tribunals are specific to tribal conflicts while tribal mediators, elder's councils, and alliances are appropriate to both clan and tribal levels. Similarly, informal sanctions are usually more effective at the individual or section levels, while alliances, or their annulment, are more effective at the clan and tribal levels. Some preventive mechanisms are appropriate to all levels, such as land use planning and education.

To conflict resolution, the most visible form of conflict management, must be added the importance of focusing on conflict prevention to diffuse disputes. In Karamoja, traditional and modern conflict prevention mechanisms – such as informal sanctions, spontaneous adaptations, alliances, land use planning, regional development and education – were and can be used. The Karimajong are experimenting with new forms of conflict resolution, some of which, particularly vigilantes, have proved quite effective in the short run, but may potentially be problematic in the long run. New mechanisms and institutional structures must be developed if the gains from successful conflict resolution are to be maintained and sustained into the future. Some examples of possible new institutions for preventing conflicts in Karamoja are: inter-tribal ad hoc committees for land use planning and resource negotiation/allocation; vocational schools to attract young warriors and the 'professionalisation' of herding (greater inputs, greater prestige, 'cowboy' competitions, etc.); and flexible herders' associations for natural resource management.

The choice of institution will depend on several factors:

- what institutions already exist, and the degree to which they are fully or partially viable;
- what mix of responsibilities are intended for the institution;

- how well the institution is 'nested' or integrated within the overall customary institutional framework of the society;
- how well the customary institution can relate or coordinate with non-customary structures.

The viability of a customary institution can be judged according to whether it functions efficiently and/or equitably, whether its structure is complete (are all positions filled? are key actors missing?), and whether its structure and function are able to remain flexible and adapt to emerging situations. The choice of a viable customary institution will also depend on which responsibilities it is expected to undertake. Informal conflict management may require informal, ad hoc institutions at the lowest socio-geographical level (e.g. villages), while formal conflict resolution between two tribes will require an intertribal institution. Choosing a customary institution, and delegating authority to it is not enough. Establishing how that institution will interact with other 'higher' and 'lower' institutions within the customary hierarchy, as well as with government and NGO institutions, is also important. Duplication of efforts must be curtailed. Institutions at different levels must coordinate to act on issues that affect several levels at the same time. Higher institutions must confer legitimacy on lower ones, and vice versa. Procedural guidelines must be developed to refer disputes to 'higher' institutions with greater moral authority.

The choice of institution will be more dynamic than that implied by a list of factors. The internal socio-political processes, the degree to which full stakeholder participation is forthcoming, and existing historical trends will determine the choice and evolution of appropriate institutions. A series of evolving and progressive, ad hoc (or transient) institutions may be necessary to allow flexibility for experimentation with different institutional solutions, rather than an insistence on creating formalised permanent structures at the outset. The latter are more suited to and liked by formal government institutions, but they may be less effective, because they lack flexibility and because they can become tools in the hands of the different political factions.

Maintaining and reinforcing conflict management mechanisms in Karamoja, an integral part of strengthening the society's social capital, can partly happen through explicitly planned measures to create or strengthen adaptive institutions, to train neutral mediators and judges, to establish mechanisms for faster and more efficient negotiation and compromise, and through development planning and implementation conducive to pastoral systems. But it will mostly occur through trial and error, through the government ensuring an 'even playing field', and through a willingness by local leaders and general public for arriving at a consensus on appropriate solutions. Local actors hope that the very process of conflict resolution will stimulate the emergence of the proper conditions for participatory planning and development. Karamoja's case is still evolving and should be watched closely. It could provide important lessons for conflict management elsewhere in Africa.

This chapter has focused on the prevention and resolution of conflicts within Karamoja. However, international conflicts continue to affect Karamoja's crisis. Cross-border disputes, civil war in Sudan, rebellion in northern Uganda, and arms proliferation will continue to impact Karamoja. The success of innovative conflict resolution mechanisms in Karamoja will depend to what extent the region can be buffered from these external conflicts.

Notes

1. Tracking is the process or method by which the herder balances the need of his livestock with the productivity of the pasture, within a continuous feedback loop based on daily monitoring (Niamir 1997).
2. A contributing factor has been that soldiers of war, under Amin and Obote, brought their ideas of independence back to the tribe (Novelli 1988).
3. 'Unearthing the Pen' refers to a ritual cancellation of a curse that was placed on 'the pen' after the Colonial Administration's punitive and forceful campaigns to educate the Karimajong.
4. Tsetse Control Department Reports, 1954.
5. 'Inclusive rights' to land are based on the recognition of a tiered priority system of property rights. For example, customary owners have first priority, neighbours have second priority, and occasional users (e.g. during droughts) have third priority (Niamir 1997).
6. The Karimajong classify people in three groups. 'Friends' are those who help the Karimajong attain his pastoral goals; the occasional allies who allow him to graze his herd in their territory, or who join him in raids against other groups.... Enemies are those who oppose the attainment of those goals.... All others [are] those who neither help nor hinder, are looked on with indifference which borders on contempt, because they have no livestock' (Novelli 1988).

3
The Social Context of Environmental Education: The Case of the Amboseli Ecosystem, Kajiado, Kenya

Ngeta Kabiri

1. Introduction

Approaches towards the conservation of the environment range from the traditional authoritarian (paternalistic) approach to community-based approaches. In between lie questions of whether the underlying philosophy should be couched in terms of moralism or materialism. Exponents of the former contend that an intrinsic value is to be gained by conserving the environment, quite apart from the material benefits that might accrue. Conservation must also be premised on the spiritual satisfaction that is derived from the connection with the universe (Kellert 1996: 209, 217). But the philosophy of conservation is probably not the most critical agony of the environmental educationists. Rather, it is the methodology of that education. How do people participate in environmental projects?

Jules Pretty has shown the need for a careful consideration of the question of how people participate in development projects (Pretty 1994). People's involvement in development projects may assume various dimensions. They may participate passively, or through giving information and/or through cosultation. In such cases, their views are not binding on the development facilitators. People may also participate for material incentives. These types of participation are least helpful in ensuring the survival of development projects. Pretty's typology (see Table 1.1 above, p. 3) suggests that the chances of survival of such projects can be improved if participation is functional, interactive and/or self-initiated. In these latter types of participation, the relationship between the people and the development agents is characterised by dialogue. In dialogical approaches, the people (those being edu-cated) participate in the process of knowing. They are, therefore, able to internalise the 'word' and truly possess the knowledge, with all that is implied by this, such as the locus of control of responsibility, accountability, quality and relevance (Freire 1970, 1973; Wood 1994).

When we move from the drawing board to ground-level implemen-tation, however, questions are raised about the participatory approach to rural development. Talk of 'involving the people' has become almost a trademark of all development agents. Organisations are constantly being urged to recognise human agency as the key to the success of their programmes (Wunder 1996: 4–5). But to talk of the local level is easy; pointing out where the local actually starts and ends is much more difficult.[1] Once on the ground, there is a galaxy of categories of stakeholders and interest groups (Lindblade 1996; Wunder 1996). What is actually meant by the 'people' then becomes highly pertinent, as well as trying to decide who are the leaders in pragmatic as opposed to jural terms.[2]

These reflections are crucial for environmental education especially, because an application of the participatory rural appraisal method immediately raises the question of whether the issues at stake are edu-cational, technical, political or ethical (Berger 1993: 103; Kellert 1996). Herein lies the need for a social context analysis of conservation endeavours even within an approach that is supposedly participatory. A social context analysis can help in approximating a meaningful par-ticipatory approach, by involving the 'people' from the stage of prob-lem(s) identification and analysis, to implementation of the solution(s) and local capacity-building. The 'people' themselves are particularly adept at unravelling the intricacies of their hopes, fears and aspirations. In this chapter I shall discuss these issues with specific reference to the Maasai and the Amboseli National Park.

2. The Maasai and Amboseli

The history of the origins of the Maasai is a matter of conjecture. Oral sources indicate that they entered Kenya from the north, from some-where in the Sudan, in or around the seventeenth century. They occu-pied the highlands of Kenya for many years, but the coming of British colonialism halted their expansionist tendencies. Through a series of land alienation pacts with the British they lost most of the land from Tanzania to the northern Rift Valley in Kenya, hitherto at their dis-posal, and became confined to their present abode, the southern region of Kenya. They occupy three administrative districts: Narok, Trans-Mara and Kajiado.

The Maasai live in homesteads, which are large enclosures surrounded by a circular thorn-bush fence. Each homestead harbours several families and their herds of cattle, sheep, goats and, occasionally, a donkey. As part of the transhumance character of the community, some members of the camp may occasionally move away to smaller camps. Modernisation is, however, changing this setting. Individuals are now moving into one-family homesteads, especially in areas where the land is being demarcated into individual ranches.

The Maasai are still among the least western-educated people in Kenya. Their literacy rates have, however, been rising, but it is difficult to quantify this since the rate of immigration of people from literate communities has also been increasing. In 1969, the Maasai constituted 69 per cent of the total population of Kajiado district. This dropped to 63 per cent in 1979, and by the 1989 census it had fallen to 57 per cent (Republic of Kenya 1989; 1990: 35; 1994: 15). By the 1989 census, of a total population of about 250,000 only 55 per cent of the population could read and write, and the district had only a total of 887 university graduates. These figures include even the immigrant non-Maasai. In 1988, for example, there were four Youth Polytechnics in this district with 181 students. Only 10 per cent were Maasai (Rutten 1992: 142). The attitude towards education is now changing positively, following the collapse of their economy.

The Maasai are traditionally pastoralists. Their economy largely revolved around cattle, goats and sheep; a few donkeys were kept for purposes of transport. Their staple diet consisted of milk, meat and blood (Kituyi 1990; Rutten 1992). The Maasai economy has been transformed by the structural dynamics emanating from the creation of the Kenyan state since the onset of colonialism. These dynamics have affected the land use practices on which the Maasai pastoral lifestyle was based, making it increasingly difficult for the Maasai to practise pastoralism, as was the case before. Initially, even as they lost land through land alienation, they kept plots of undivided land to graze their cattle. With time, land ownership has become individualised, so that cattle movements in search of pasture and water are increasingly being curtailed. Land individualisation implies excluding certain vital areas from communal pasturing, areas that traditionally served as a drawback recourse during drought. Hence, in the event of drought, for example, there will be no escape routes, which will spell doom to the livestock herders. Consequently, agriculture is now becoming the only other mode of production into which the ordinary Maasai might venture (Rutten 1992: 457). But this option is a problematic one, given the nature of their land.

The general topography is one of plains and occasional volcanic hills and valleys. The soil types consist of shallow, poorly drained heavy clays, which are often waterlogged in the rainy season and very heavy to work in the dry season. The volcanic hills consist of either rock with little or no soil cover, or of highly porous, erodable ash. Estimates suggest that only 1 per cent of Kajiado's entire territory is high potential, while 98 per cent is low potential (Republic of Kenya 1990, 1994; Sindiga 1981). The drainage and water resources are mostly a characteristic of seasonal streams and rivers, with only one perennial watercourse, the Embakasi river. Rivers derived from the snowmelt on Mount Kilimanjaro seep underground in the volcanic rocks. The underground flows emerge as springs and ultimately form rivers that have perennial flows in their upper reaches, but reduce to negligible quantities in their middle and lower reaches. They then disappear underground within a few metres' flow of their hill sources. Thus they are not available in the low-lying grazing lands where they are needed. Some permanent swamps exist, along with seasonal lakes such as Lake Amboseli, which floods in wet season and dries up in dry periods. The rainfall is insufficient, fluctuates in its annual totals and there is a high evapotranspiration rate (Republic of Kenya 1990, 1994; Sindiga 1981).

The main vegetation type consists of wooded grasslands, open grassland, semi-desert bushland and scrub. The dominant plant species are acacia trees and grass species (Republic of Kenya 1990, 1994). Vegetation cover has socio-cultural significance for the Maasai. Forests, for instance, are the pharmaceutical stores for the society. Grass is used not only for feeding cattle, but also for building houses. The vegetation cover is also home to a substantial number of wildlife. Wildlife was appropriated for social significance in various ways: buffalo hide made shields; rhinoceros horns made containers for tobacco; giraffe tails made fly whisks for elders; ostrich feathers made head-gear for warriors; and hide and skin from revered animals such as the lion and colobus monkey made robes of honour for the elders. The Maasai, however, did not appropriate game animals for human consumption. All these uses that were made of the wildlife have now been alienated from the Maasai because the government has taken control of the wildlife sector to develop the tourist industry. This alienation assumed institutional dimensions when the Amboseli game park was created.

The park was formed out of Maasailand during the colonial period, when it was managed by the local county council. This status changed in 1974 when it was brought under national management. Its size was reduced to 240 sq. km and it was renamed Amboseli National Park. Today, it is run by the Kenya Wildlife Service. The park's habitat includes the seasonal Lake Amboseli, hardpan alkaline plains, and various woodlands, including acacia; swamps and bushes. The best watering places are inside the park boundaries. The dominant animal species are elephants, wildebeests and plain gazelles (Republic of Kenya 1990: 8). The park and issues adjunct to it have been a theatre of contestations between the Maasai and the colonial and postcolonial governments. The Maasai have largely been sidelined, or involved only marginally in the management of the park resources. This exclusion and lukewarm Maasai involvement has triggered a train of events which has involved an increase in the pressure on wildlife and ecological degradation.

The wildlife seems to have been caught up in a circle. When, for instance, there is no poaching, they move about more extensively, thereby increasing competition for grazing and wreaking havoc on farms, livestock and people. This triggers a retaliation, which leads to their confinement to the park. This confinement in turn leads to desertification. The resulting absence of trees drives out the grazers and as they trudge outside the park, they create conflicts in which they ultimately end as the losers (Gazette 1992; Pearce 1998). In this study I elaborate on the role of the community as an interested party in natural resource management and assess community participation by considering the empirical background of the key issues at stake in Amboseli.

3. Land use practices

The Amboseli ecosystem is presently beset with a crisis of degradation due to competing land use practices. In what follows I review these practices, inquire into the degradation crisis they have engendered, and raise the possibility of a community response to this crisis, to show the importance of a social context analysis.

3.1 Agriculture

While the Maasai are not traditionally a farming community, a few of them undertook limited farming activities (Bekule *et al.* 1991). Recently, involvement in agriculture has increased, because the subdivision of land has rendered livestock raising difficult; drought has devastated the livestock enterprise, thereby necessitating a need for agriculture as a weapon of rebuilding their herds and thus containing impoverishment; and the commercialisation of agriculture has destigmatised farming (Bekule *et al.* 1991; Kituyi 1990). There are three discernible agricultural types.

3.1.1 Large-scale commercial farming

This is mostly industrial-oriented and the crops are destined for the export markets (Campbell 1993: 265). The practitioners are both Maasai and non-Maasai who have acquired lands that are better suited for farming.

3.1.2 Medium-scale agriculture

This involves sharecropping between Maasai landowners and non-Maasai immigrants. A study of one of these horticultural societies showed that only about 16 per cent of the cultivators were Maasai (Berger 1993: 27, 104). The land leases in these partnerships are not always legal and hence evictions are not rare (Rutten 1992: 315). Such evictions have adverse implications for sustainable agriculture; it would be expected that the tillers are ruthless in their exploitation of the land with little, if any, investment in the reproduction of the land for the future.

3.1.3 Small-scale agriculture

This is mostly practised by poverty-stricken common Maasai. Having been pushed from the more arable lands by the above two, they end up in marginal lands with little agricultural potential. They seek cultivable lands in the wetland areas near streams, around mountains and swamps – areas traditionally reserved for dry season grazing. Consequently, the average Maasai response to agriculture is 'hostile'. This hostility is manifested in such responses as cutting down fences and setting farms on fire (Kituyi 1990: 103, 106); during drought, farms are occasionally treated as pasture lands (Nation Group of Newspapers 1997a). This combination of land use practices has precipitated the environmental crisis associated with the Amboseli ecosystem (Western 1969).

3.2 Agricultural practices

The various features of these practices include chemical and fertiliser applications, cutting down trees and a general clearing of land of all vegetation and ploughing, which is sometimes too deep and short of fallow periods. But the most critical land use practice seems to be irrigation. Irrigation is concentrated primarily in the Loitokitok region where Amboseli is located (Rutten 1992). Irrigation will probably gain momentum once the road network improves, thereby favouring horticulture. Irrigation activities divert springs and rivers at their source thereby depriving people, livestock and wildlife downstream (Berger 1993: 22). The irrigated schemes are in the lower rangelands, once the dry season refuge for livestock and migratory wildlife (ibid.: 34). In the 1980s, the government intervened to provide most agricultural and irrigation services (Kituyi 1990: 95–6) thus exhorting the Maasai to settle into a sedentary life. The agricultural phenomenon has had far-reaching ecological consequences.

3.3 Implications of agricultural practices

3.3.1 Ecological degradation

The upswing in farming activities largely accounts for the environmental deterioration now confronting this region. Agricultural activities have affected rivers and the capacity of swamps, with some disappearing. This is reducing the holding capacity of the rangelands (Campbell 1993: 265; Rutten 1992: 315). The Namelok swamp adjacent to Amboseli, for example, was adjudicated and granted to group ranches. Thereafter, the swamps were drained for farming and the wildlife was driven out. By 1980, the game had all but disappeared from Namelok (Western 1994: 29). Woodlands are also being denuded as trees are cut down, not only to clear the fields but also to service the requirements of a sedentary life: the provision of charcoal, fencing posts, firewood, etc. (Kituyi 1990: 106). Prior to sedentarisation, land recovery could take place because (unlike the permanent settlements) the nomadic way of life would cause the abandoning of certain places for a time. The slopes of Mount Kilimanjaro, for example, were a well-forested zone in the 1970s; it was then sold to non-Maasai, and now it has become smallholder agricultural land with remnant forest patches (Berger 1993: 32). The loss of ground cover may be leading to desertification (Berger 1993: 143). Also, about 90 per cent of the park's full-grown acacia xanthogholea trees have died since the early 1980s because of a rise in the water table and increased salinity due to felling of trees on the slopes of Mount Kilimanjaro.

The denudation of woodlands has reduced the water available from the springs which originate in the plains, and has also increased the risks of landslides due to diminished water-conserving capacity. Further, soil erosion has increased, partly as a consequence of the foregoing and also of agricultural practices that encourage run-off and soil loss, as well as salination of irrigated fields (Rutten 1992: 315).

3.3.2   Fencing (and wildlife viability)

The need to protect crops from wildlife has led to fencing, which inter-feres with wet season dispersal of wildlife, thereby undermining the viability of the parks and game reserves. The spectre of fencing lies in the fact that it interrupts migration cycles, in turn leading to over-crowding of wildlife and eventual land decimation. Thus fencing is a threat to the ecological integrity and naturalness of the parks (which is what most tourists value) due to its insularisation effects (Rutten 1992: 323–4, 352, 368; Campbell 1993: 268; Western 1994: 42, 49). The over-population of elephants in the parks, partly due to the demise of their seasonal migration, may also severely reduce Amboseli's biological diversity (Western 1994: 50). But the response by the conservationists towards fencing does not seem to have been unanimous. While some have opposed it, others have been for (and against?) it. Exponents of fencing have even been willing to help farmers put up electric fences (Berger 1993: 99–100; Western 1994: 42). The European Union, for example, provided money for fencing wildlife sanctuaries (Nation Group of Newspapers 1997b). A 1994 political campaign against the then Director of the Kenya Wildlife Service (KWS) vilified him for being discriminative in assisting the people in their fencing endeavours (Nation Group of Newspapers 1994b), while the probe committee inquiring into his performance criticised him for his drive to fence off the parks and game reserves completely (Nation Group of Newspapers 1994c).

But on the whole, KWS does not favour fencing. Western refers to how fencing is being used by some Group Ranches as a weapon for negotiating with the KWS over revenue-sharing. These ranchers threaten to fence their lands if they are not paid wildlife utilisation fees. The fact that they are among the first to be paid (assuming it is because of these threats, and Western seems to suggest so) (Western 1994: 43) shows that KWS discourages fencing. Indeed, the KWS has threatened it will stop paying wildlife utilisation fees to landholders if fencing is introduced. The government is trying to increase the wildlife utilisation fee so as to persuade the Maasai against fencing or restrict-ing wildlife (Campbell 1993: 268). Since agriculture cannot be expected to coexist with wildlife without the medium of a fence, an attack on fencing amounts to calling on the Maasai to abandon agriculture.

On the other hand, implicit in the support for fencing is the question of whether it can at the same time be argued that agriculture should be abandoned. A failure to resolve the fencing/agriculture nexus would imply an increase in the pressure on wildlife.

3.3.3 Intensification of conflict with the wildlife sector

When animals roam outside the parks, crops are destroyed and they compete with livestock and farmers for water, especially when agriculture involves irrigation. The animal menace escalates beyond effects on livestock (predation, pasture and diseases) and the killing of people. Animals such as elephants, warthogs, monkeys, gazelles, waterbucks, zebras, bush backs, and so on (Berger 1993: 42, 109 n. 2), previously considered harmless, are now victimised because they ravage crops. In some areas, more than half the crop, it is estimated, is destroyed by herbivores (Western 1994: 41). Attitudes towards the animals have changed as a result. Wildlife that destroys crops are killed, particularly given the abolition of compensation to farmers for crop damaged by wildlife (Nation Group of Newspapers 1994a). As recently as November 1997, the Director of the KWS was still campaigning for support among donors on the platform of human–wildlife conflict (Nation Group of Newspapers 1997c). Agriculture is largely at the centre of this conflict.

3.3.4 Conflict with livestock owners

Conflict is caused by several issues. One is pollution, which affects human and animal health through contamination downstream by the use of chemicals and fertilisers. Irrigation has led to the loss of dry season grazing areas, which in turn pushed the livestock keepers to drier parts. This loss has implied pressure on existing grazing areas thereby causing degradation of vegetation cover (Matampash 1993: 37; Republic of Kenya 1994: 19; Western 1969). The Maasai believe that agricultural activities are leading to reduced water availability through the diversion of water from swamps for irrigation (Berger 1993: 42, 68, 96–7, 103–4, 141; Rutten 1992: 95, 315). Moreover, when cultivation is done along rivers and swamps, the watering points become blocked and the animals have to travel long distances, a situation that the Maasai resent (Berger 1993: 103–4). Maasai landlords side with the immigrants against cattle keepers, with implications for the homogeneity of the Maasai community as a unit of analysis for environmental education. (Can Maasai interests still be discussed as if they are generally homogeneous in terms of attitude to resource use? How many interest groups are likely to be there – the livestock keepers, Maasai cultivators and Maasai landlords?)

3.3.5   *The livestock–wildlife sectors*

In the past, the wildlife was seen as in competition with livestock for land. Maasai culture is presented as having stood for the coexistence of wildlife and people (Kipury 1983). The dichotomy was introduced by the state's creation of wildlife protected areas which alienated the Maasai from their land resources (best rangelands and water reservoirs). Yet the largest proportion of animals (60–80 per cent) continued living outside these protected areas, thereby competing with the Maasai over the land left to them (Matampash 1993: 37). The crisis has been exacerbated because in spite of the hefty income to the state by the wildlife industry, the Maasai, who consider themselves as the wildlife custodians, receive only a paltry share, if anything (Republic of Kenya 1990: 78). As a result, the Maasai have developed grievances against the animals. The wildlife are now viewed as direct competitors for grazing and watering; this becomes particularly acute as more and more land is alienated for agriculture. Moreover, wildlife are predators for livestock (Republic of Kenya 1994: 20). Both wildlife and livestock are accused of spreading diseases. The calving wildebeest is held as the cause of malignant catarrh; it acts as a reservoir for blood parasites that cause trypanosomiasis in cattle (Republic of Kenya 1990: 68; 1994: 20; Rutten 1992: 102, 368–9; Sindiga 1981: 70). On the other hand, some conservationists see the livestock as spreading diseases. Cattle grazing is associated with rinderpest, which decimates substantial herds of wildlife (Nation Group of Newspapers 1997c). Conservationists have also complained that when the livestock biomass reaches a high percentage, it damages the environment – though sometimes evidence does not bear this out (Berger 1993: 39, 43–4 n. 39). Western (1969), however, suggests that a high livestock population, and a grazing sequence that goes against that of the rest of the animal communities, clears the available grass such that the herbivores are left with the acacia only, which they then decimate. In terms of economic returns, some conservationists argue that a greater mileage can be gained from wildlife than from livestock (Mitchell 1969; Western 1969). Some have, all the same, argued that with adept livestock management, conflict between livestock and wildlife is not inevitable (Mitchell 1969).

4. What can be done?

The three sectors – agriculture, livestock and wildlife – are currently in conflict, but are they really mutually exclusive? The wildlife sector is a major income-earner for the country and the hardest to eliminate, hence its primacy. The question of where to look for a solution seems likely to be directed at agriculture and livestock, with the former being the most vulnerable, because agriculture does not enjoy grassroots support, being largely a phenomenon of the non-Maasai and some Maasai elite. The possibility of grassroots subterfuge (burning of, and grazing on, the farms), can always render its elite connection inconsequential. Several suggestions have been made to modify the effects of agriculture.

4.1 Buffer zones

These can be set between parks and individual ranches. The Maasai, however, object to this solution due to the issues of ownership, compensation and the distribution of revenues accruing from them (Campbell 1993: 268). The problem is that they do not trust the government to keep an agreement: a history of betrayal makes them circumspect. This mistrust is possibly fuelled by media reports that the conflicts within certain government circles have partly to do with the struggle over the private (mis)appropriation of land gazetted as wildlife parks (Nation Group of Newspapers 1994a,d). As recently as 1997, sections of the Maasai were contesting the appropriation by individuals of land they had ceded to the government for a livestock improvement scheme. Apparently, this scheme failed and individuals apportioned the land to themselves. The Maasai now contend that this land should serve as a wildlife corridor (Karengata 1997). They are unlikely to part with even an inch of their land again (Western 1994: 29; McKinley 1996).

4.2 Legal action

This may include banning deforestation, squatting and prohibiting agricultural activities too close to watercourses, and has been attempted since the colonial period, but has yet to pay dividends consistently (Rutten 1992: 213–14). The framework, however, seems to work for the protection of animals when the local community cooperates against poachers, related animal threats or even deforestation (Berger 1993: 140–1; Western 1994: 46). If certain measures can only work with a community anchor to them, this is a statement of the inevitability of community goodwill and, by extension, community participation in conservation endeavours.

4.3   Agro-forestry

This could entail part or a combination of activities such as 'plant by cultivation' schemes where people may be allowed to stay in forest areas and cultivate as they plant trees. Such trees could be multipurpose: fruit-bearing, fodder and fuelwood supplying. Soil conservation activities could be undertaken, involving the digging of cut-off drains, construction of terraces and the planting of trees and grass. Similarly, farming techniques could be introduced to combat soil erosion, such as crop rotation and alternate cropping, water-spreading and lay farming (which entails periods of arable cropping followed by one of grass and legumes). The implications for livestock production would be to develop an agro-livestock sector, yet these two sectors may not be able to occur simultaneously with a third one, the wildlife sector. Agricultural activities have been found to exert pressure on natural resources, particularly if the expansion is into classified forests (Lisa *et al.* 1996).

5. Are there critical junctures, which call for educational intervention(s)?

The need for educational interventions, given the foregoing, cannot be overstated. Definite processes of degradation, bordering on desertification, call for the deployment of ecologists and managers into the ecosystem. The conservation strategy and management of biodiversity of the region must be addressed. As I have suggested, the Maasai ought to be an integral part of these initiatives, and even the government has admitted to the lack of comprehensive joint community participation policies (Nation Group of Newspapers 1994c,e). Heavy expenditure is also incurred in resolving wildlife/human conflict outside parks; yet in these areas the KWS raises no income (Nation Group of Newspapers 1997c). To this extent, the KWS policy already revolves around mitigating wildlife/human conflict (Nation Group of Newspapers 1997d). To be successful, involving the Maasai will be of paramount importance.

Observers, including the Maasai themselves, have argued for a more widespread extension education programme and community mobilisation for the purposes of reversing the 'present observable trends' in the degradation of the environment (Berger 1993: 144; Matampash 1993: 38, 44). That the community could be a viable option is suggested by the attempts – already functioning at various local levels – to provide group environmental protection around swamps, river banks and springs. Communities are also helping to apprehend culprits whose activities

degrade the environment. Conservation activities such as the use of fur-
rows in irrigation areas, agroforestry, and so on have also been
attempted. Elsewhere in Maasailand, associations are applying cultural
and scientific programmes to wildlife management and conservation
(Nation Group of Newspapers 1994a). In 1996, some Maasai living next
to Amboseli opened the first community wildlife reserve (McKinley
1996). These approaches suggest that it is possible for the situation to be
salvaged, one way of doing so being through environmental education.
But can the critical actors, the Maasai, be relied on to follow the task
through to its logical conclusion? A study of their interaction with the
environment in the past argues for an answer in the affirmative.

6. Maasai and traditional interaction with their environment

Studies of the Maasai past suggest that there were conscious attempts
at environmental management as evidenced by, for instance, the
elders' regulation of grazing patterns such that degradation through
overgrazing was unknown (Graham 1989; Matampash 1993: 31).
Livestock movement was geared towards avoiding grass damage during
critical times (Berger 1993: 18). Grazing took place on lowlands during
wet weather, when the water supplies were good, and then shifted to
higher and wetter regions during the dry season, conserving water and
grazing resources, and allowing for fertility restoration (Tignor 1976: 38).
Social mechanisms such as rebuke were directed towards those not
adhering to these norms. Moderate burning of grasslands controlled
ticks and diseases and promoted growth of nutritious grasses (Berger
1993: 18).

The Maasai are said to have considered wildlife as their second cattle,
because during drought, when their herds had been depleted by the
vagaries of nature, they would resort to wildlife for sustenance
(Western 1994: 20). Other sources, however, contend that the Maasai
did not hunt wildlife for meat and despised those who did (Kipury
1983). The Maasai are also said to have been sensitive to animal mistreatment, seeing it as indecent to subject livestock to drudgery, for
example, such as the use of ox for ploughing.

Pastoralism was viewed as part of creation, in contradistinction to
farming, which was portrayed as a curse and an abuse of Mother Earth
(Kituyi 1990; Matampash 1993: 36). The Maasai social life was said to
be interwoven with the wildlife environment, and they derived a measure of spirituality from the environment. Myths, legends and tales

about the land and environment were narrated in a sanctified manner (Kipury 1983). Land, for instance, was viewed as God-given, and thus as communal property, and its aridity was seen as a punishment by the Creator, a sign of annoyance for the destruction of the environment (Matampash 1993: 31, 35–6). Animal names were used to symbolise honour and horror. The animals of a greedy person were stigmatised and said to bring diseases (Kipury 1983; Kituyi 1990: 170–1). Certain wildlife species were associated with ceremonies and medicine, while other species, such as Columbus monkeys and the ostrich, were said to have cultural significance (Berger 1993: 105; Kipury 1983; Matampash 1993: 35–6). Positive Maasai relationships to wildlife can therefore be developed on the basis of a cultural legacy, and their participation in environmental conservation can be promoted.

7. Maasai perceptions of the current environmental situation

In spite of this traditional legacy, have the Maasai lost touch with a conservation ethos? This does not seem to be the case, the myriad of environmental predicaments besetting this region notwithstanding. The Maasai are still aware of the issues at stake. A survey by Berger (1993: 103–6) and Matampash (1993: 39, 41) reveals a profound awareness among the Maasai of environmental and resource problems – such as overgrazing, agriculture and the disappearance of wildlife species – besetting their region. But they nevertheless continue to maximise their stock since there seems to be no other way of securing a living. Moreover, they consider overstocking as having been imposed on them through land alienation (Matampash 1993: 39). The tragedy in the wildlife sector is largely seen in the context of their failure to benefit from it, not to mention the host of betrayed promises from the government. The history of Maasai relations with the establishment suggests that the climate for participatory conservation has been frequently cloudy (Western 1994). The legacy of these strained relations is that the dwindling fortunes of the ecosystem are directly linked to Maasai perceptions of the few benefits that accrue from this ecosystem. When their interests have been in the forefront, things have flourished well, while during periods of their mistreatment (as they perceive it) chaos has reigned. Thus the history of conservation efforts suggests that a successful conservation approach would have to include considerations of their interests; hence the need for a social context analysis.

8. History of participatory conservation

The history of participatory conservation dates back to the period of land alienation for the creation of animal sanctuaries, which separated the people from the wildlife. Though this has its genesis in the colonial period, in the 1940s, initially no great problem emerged, since pressure on land due to alienation was still mild and agriculture was not extensive. Even so, as early as the 1950s, echoes of the community benefiting from the wildlife were heard. The director of the National Parks and the warden of Amboseli, for instance, proposed that the Maasai get a share of the revenue from the parks. They received no support from the government, however (Berger 1993: 38–9). Thus government intransigence is as old as the notion of benefit-sharing. In the 1960s, conservationists pressed to exclude the Maasai from the parks, with respect to grazing. In 1968, an offer to provide water resources outside the parks in exchange for Maasai exclusion was rejected by the Maasai since they had not reaped any benefits from the parks (Western 1994: 25). In 1969, a Maasai park plan was briefly accepted and then rejected because of the suspicion that it would result in land alienation, and the Maasai were not prepared to cede their land. In 1971, the failure of the local plans at managing the parks forced the government to annex certain land for the parks. This triggered protest spearing of wildlife until the Maasai won back certain sections of the annexed land (Western 1994: 29). In the 1970s, the government pursued a Maasai-sensitive policy which promoted communal benefits from the parks, leading to a climate of toleration between humans and wildlife. Poaching and spearing became rare, resulting in the recovery of elephant and rhino populations (Berger 1993: 12; Western 1994: 32–7) which had dropped from 1,000 to 480 rhinoceros and from 75 to 8 elephants between 1967 and 1977 (Western 1994: 46). This equilibrium was, however, distorted in the 1980s when the government largely reneged on its obligations to the Maasai community. The Maasai not only dropped the spirit of cooperation, but they also become anti-conservation in a tone characterised by poacher-friendliness, which marked both an increase in spearing of, and a dwindling in, elephant and rhino populations (Berger 1993: 39; Western 1994: 39–40).

Attempts were made in the 1990s to re-enact the 1980s conditions, but inertia remained on the part of the government. This led the Maasai to pursue a carrot-and-stick policy, and the fate of the ecosystem largely hung in the balance. But of most interest in this drama has been the Maasai tendency to show conservation sensitivity even in the face of government procrastination. Since it dawned on the Maasai that they could reap benefits from the ecosystem, particularly in relation to tourism, they have been keen on preserving this possibility, government betrayal notwithstanding. They have, for instance, opposed Group Ranch subdivisions due to its implied havoc to wildlife migratory routes, though at the same time threatening to put up fences if they do not receive a share of revenues. Thus the ecosystem has largely remained open, with viable migrations and a wildlife increase. The ranches have also formulated plans to protect wildlife outside the parks. Hence, while other ecosystems in Kenya have had problematic experiences, reflected in wildlife decline and negative community attitudes, the Amboseli ecosystem has shown tendencies towards resilience (Berger 1993: 140; Nation Group of Newspapers 1994f; Western 1994: 42–6). How far this fluctuating situation can go is not certain. While much of the future, in light of the current events, is unpredictable, the havoc that would result from an exclusion of people from assisting in management of the ecosystem is obvious, hence the need for a community participatory approach.

9. The lessons of the past

Even the KWS has acknowledged that community participation is crucial if the Amboseli ecosystem is to be maintained. Its policy is geared towards community service, getting people to take an active part in conservation (Nation Group of Newspapers 1997c; Weekly Review 1997). But there are dissenting voices. Critics question the viability of this approach in terms of the benefits that accrue to animals and people. A former Director of the KWS, despite being an exponent of the community sharing the benefits accruing from the parks, had problems with how this should be shared (Nation Group of Newspapers 1994g). Concerns also remain about leaving the sanctuaries to incompetent politicians at the local level (Nation Group of Newspapers 1997e; Weekly Review 1997). The state includes those opposed to landowners taking control of wildlife sanctuaries (Nation Group of Newspapers 1997d). State functionaries question the forming of community associations to manage the wildlife resources. They perceive them as political groups and thus feel threatened by them (Graham 1995). Thus while the past does not seem to have taught the state that community-based conservation programmes are the only viable alternative, it teaches the Maasai that they can benefit only if they are an active and integral part of managing the ecosystem.

10. The Maasai as future actors in the conservation project

Some suggest that the Maasai have developed a sense of apathy to environmental imperatives since they no longer control their land (Matampash 1993: 38–9; McKinley 1996). The critical issue here is that of representation in matters that demand the goodwill of the Maasai: do the Maasai have common interests in the management of the ecosystem? In the past, the centre of power lay with the elders and their position was binding on the rank and file (Sankan 1971). This trend is changing; the centre of power today seems to oscillate between the diminishing power of the elders and the emerging influence of a new elite (Kituyi 1990: 207). The elite are not themselves a homogeneous class, as were the elders, and their interests in the community are suspect. They are seen as more aligned with the national elite who pursue state policies not always favourable to the Maasai; not to mention their appetite for enriching themselves at the expense of the Maasai masses (Berger 1993: 21–2; Campbell 1993; Graham 1989; Kituyi 1990). This has led some to describe an incapacitation of traditional leadership without a provision of a workable substitute (Bekule *et al.* 1991). As mentioned above, tension has been introduced by the migrant farmers who are benefiting a section of the landed Maasai and thus setting these against the rest, who, on the whole, are resentful of the farming sector. In addition, alliances are rooted in a history of subsection/clan rivalries, now made worse by the formation of group ranches and manifest in disputes over boundaries and resource access (Bekule *et al.* 1991; Berger 1993: 116). The question of resource accessibility has also caused a cleavage between the political elite and those Maasai organised within the civil society (Nation Group of Newspapers 1994a).

But even if the Maasai were to emerge from their factions, their unified position would not translate automatically to cooperation with the conservation sector. If anything, recent developments suggest that the unanimous voice of the past could be replayed, with the possibility that they could submit demands not always easy to meet, such as the taking over of almost the entire control of the parks. Some have argued that the wildlife is Maasai heritage, and not national heritage; others have threatened to kill animals if Maasai interests continued to be ignored (Nation Group of Newspapers 1994a). The current political temperatures generated by their leaders vis-à-vis their position within the national political economy is one of marginalisation, which they are seeking to end. This could be translated to laying claims over the natural resources within their lands (Nation Group of Newspapers 1994a; Okondo 1993). Thus future conservation efforts will have to deal with a situation where the theme of marginalisation is critical.

11. The discourse of marginality

What are the major highlights of this claim of marginalisation? Maasai politicians have few rivals among the minority ethnic groups in Kenya when it comes to the oratory of dispossession. Key issues here are land, poverty and exploitation.

11.1   Land

In a history of land alienation dating from colonial times, areas of high productivity have been alienated either by outsiders or the emerging Maasai elite (Berger 1993; Kituyi 1990; Matampash 1993: 31–2; Rutten 1992). The Maasai have been pushed to the marginal lands. Even here, a new land policy of subdividing remaining lands into individually owned units will dispossess nine-tenths of the people (Hillman 1994: 61; Kituyi 1990: 201). Moreover, most of those who get land find themselves confined to units too small and arid to be economically viable either for ranching or rain-fed plantations. They also face a looming ecological disaster resulting from intensive competition for scarce resources, which could lead to environmental degradation as elsewhere in Kenya's arid and semi-arid lands (Hillman 1994: 61–2).

11.2   Poverty

Kenya's arid and semi-arid lands, since the colonial era, have been characterised by poor infrastructure, when there has been any at all. They are poorly integrated with the rest of the economy. Their only individually owned resource – livestock – is frequently decimated by drought. People have to travel 40–50 km in search of pasture and water, with many of their animals dying before reaching the water points. In 1992, the government estimated poverty among pastoralists in these regions at 43 per cent (Economic Review 1997; Republic of Kenya 1997). This contrasts sharply with a colonial government report, which had placed the Maasai as one of the wealthiest people in Kenya in the 1930s (Rutten 1992: 23). The irony here is the one we referred to earlier of the Maasai being the keepers of the largest state income earner (the wildlife – and still they live in abject poverty. Moreover, individual white-collar workers and entrepreneurs working in the wildlife-tourism sector earn hefty salaries, far higher than the gross domestic product of several Maasai villages (Nation Group of Newspapers 1994c, 1997c; Republic of Kenya 1990: 78; The People 1997). Even other landowners having wildlife sanctuaries reap benefits (Nation Group of Newspapers 1997b,d). Some of their leaders have argued that the Maasai should reap more from wildlife just as other communities receive revenue from the crops they grow (Nation Group of Newspapers 1994a; Pearce 1998; The Guardian 1999). Moreover, their counterparts in Narok district control the Maasai Mara Game Reserve. Thus the question 'why not the Maasai'? provides the framework on which the issue of wildlife is being looked at. The Maasai are now treating this as pure exploitation and contempt (Nation Group of Newspapers 1994e; Richard 1993: 244; Tignor 1976).

11.3 Exploitation

The discourse here surrounds the grabbing of land and resources by non-Maasai (Nation Group of Newspapers 1997f; Human Rights Watch 1993: 42). They grieve that 'a single community have occupied Maasailand from Ngong to Loitokitok' (Nation Group of Newspapers 1997g). Against this background they lament the grabbing of their property and the 'suck[ing] of our blood' (Nation Group of Newspapers 1996), ostensibly, 'because they have been a little bit disadvantaged in education and other spheres of life' (Nation Group of Newspapers 1997g). They complain about their failure to benefit from the wildlife resource even when they are not only its custodians, but also the ones who bear the brunt of its havoc (Nation Group of Newspapers 1994a).

11.4 Liberation overtones?

The leaders of these communities seem to be on a warpath to assert their interests. A glimpse of this is provided by an incident when a Cabinet Minster from a big ethnic group and a Member of Parliament from a small community almost came to blows in the national assembly. A Cabinet Minister from the Maasai community said:

> I was happy to see the fighting, it shows the magnitude [extent to which] some people who have been historically marginalised and who have no where to go [for reprieve] can go to make themselves heard. (Nation Group of Newspapers 1996)

The same politician, while commenting on an incident of ethnic clashes where 'his' people killed members of another ethnic group, stated that he had no regrets since those people had suppressed the Maasai and 'we had to say enough is enough. I had to lead the Maasai in protecting our rights' (Watch 1993: 42).

Thus, they present themselves as having a right to fight back against those 'oppressors who have invaded us'; further to this they declare, 'the situation must change, we must begin to look at things from the right perspective' (Nation Group of Newspapers 1996). These inflammatory statements thus suggest marginalisation at a communal level. But we may be dealing with a situation where opportunistic politicians are misusing the name of the people for their selfish ends and thus being of no practical consequences on the ground and, by extension, on environmental conservation. In other words, is there a scenario, which can explode over the issue of the wildlife, as it has done at the ethnic level, and thus render conservation endeavours problematic?

12. The social scientists' view

Beyond partisan politicians, social scientists, both non-Maasai and non-Kenyan, have elaborated on the subjugation of the Maasai, thereby mitigating the misuse approach. Richard (1988) documents the misfortune that beset the Maasai, such as being sold into slavery, as a consequence of the activities of their neighbours in the pre-colonial period. Galaty claims that the Maasai have been perceived as backward by the other richer ethnic groups. He further shows that the question of political insecurity and mistrust of the postcolonial state among the Maasai dates as far back as the onset of independence (Galaty 1988: 159, 160–3). If this is so, can they now claim that their fears were prophetic? Research suggests that a study of Maasai–state relations (both colonial and postcolonial) is a study of betrayal and domination (Berger 1993; Rutten 1992; Western 1994). As Hillman observes,

> It is a well documented historical truth that the Maasai pastoralists have been persistently and variously required (by force, trickery and legalisms) to pay the price for social, economic and political decisions made by others and for the benefit of others. (Hillman 1994: 57; See also Matampash 1993: 34)

What then do these experiences suggest? Certainly, they can suggest a shift away from, and not towards, a trust in the state. The Maasai have developed a stake in what goes on at the national level. Within this context:

> As the Maasai are exposed to competition with other Kenyans, increased pressure on resources traditionally appropriated by the Maasai leads to a feeling of shared threat from the other people. (Kituyi 1990: 230)

The Maasai, as a group, do seem to have a case to lay claim to resources in their region. This makes the security of these resources a factor of how far they identify with them. Herein then lies the basis for a social context analysis in favour of formulating a natural resource management policy for the Amboseli ecosystem.

13. Summary and conclusion

My study is concerned with the question of the community in environmental conservation, and has shown that problems that may be read as educational, ethical or even technical may actually be questions of political economy. To address these problems calls for a dissection of community interests. It is our case that no ABC of natural resource management would make sense to a people who do not identify with the resources being conserved. Thus the question of the 'benefits' of these resources will go a long way in setting the pace of natural resource management in this region as we enter the twenty-first century. This is a question that is better handled through a social context analysis. This is so far one of the most reliable tools in defining the key signifiers of community participation, namely: community interests, the people, and even the pragmatic leadership of the community.

Notes

1. See, for example, Graham (1989); Matampash (1993: 40–1); Nation Group of Newspapers (1994a) on the Kajiado County Council versus the Kenya Wildlife Service and the Maasai rank and file; Vigdis *et al.* (1980) on the dispossession of Maasai women; Western (1994).
2. Those crowned as legal leaders do not necessarily move public opinion, so critical for community conservation.

Part II

Local-Level Projects Attempting to Overcome Unsupportive National Contexts

4
Addressing Livelihood Issues in Conservation-oriented Projects: A Case Study of Pulicat Lake, Tamil Nadu, India[1]

Devaki Panini

1. Introduction

In most conservation-oriented projects in India, livelihood issues are most often secondary to the goals of conservation. Current environmental discourse in India posits two divergent views: one advocates strict conservation and maintenance of the sanctity of protected areas; the other emphasises that people living in protected areas should not be alienated from these areas by a strict administrative regime. The latter view supports the vital role of local communities in effective conservation and natural resource management. This chapter, and the insights derived from the case study of World Wide Fund India's (WWFI) ecological restoration project on Pulicat Lake in south India, endorse the latter view. By providing an in-depth account of the project, and by reflecting on the inherent flaws in the project framework, the chapter attempts to bring out the importance of livelihood issues in conventional conservation or ecological restoration projects.

   The main preoccupation of this chapter is therefore to understand whether livelihood interests can be harmonised with conservation goals or whether these are really mutually exclusive and incompatible. Specifically, I try to examine whether the goals of conservation can be integrated with the livelihood needs of the main beneficiaries of any conservation-oriented project, or whether there is a fundamental conflict between these two interests. Another concern is the highly questionable practice followed by certain conservation NGOs of assuring livelihood benefits to communities in order to secure funding from international funding agencies. Conservation non-governmental

organisations (NGOs) often graft objectives of improvement of livelihood in conservation-oriented projects, merely to match the terms of
reference and priorities of funding agencies. This is disturbing because
these conservation-oriented projects rarely address the livelihood needs
of the beneficiaries, yet livelihood needs are a part of their objectives
since this is an assured formula for securing financial assistance from
international funding agencies.

2. The aims of the eco-restoration project

The observations about livelihood issues in conservation projects
are drawn from my own experience and insights gathered during
the course of project management of the WWFI's project titled *Eco-
restoration of Pulicat Lake with Fisherfolk Participation*. WWFI originally
started the project in 1995 in collaboration with the Chennai-based
Centre for Research on New International Order (CRiNIEO). The funds
for the project came under the Joint Funding Scheme of the UK
Department for International Development (DfID) and were jointly
given by the DfID and the UK World Wide Fund for Nature. The aim of
the project was the scientific restoration of the lake habitat and the
enhancement of various endangered species of fishes, crabs and prawns
in selected areas of the lake with the support and involvement of
the fisherfolk of Pulicat Lake. The scientific restoration experiments
included scientific studies through sampling and analysis of the ingres-
sion of larvae of diverse species, substratum management, silt management and mangrove plantation in selected areas of the lake. It also
involved plankton studies and recording of out-migration of breeders
to spawn in the inshore waters, and conservation and restoration of
rapidly depleting indicator species. The studies included monitoring
various ecological niches preferred by various species and biological
communities in the lake. Such experiments were expected to improve
the biodiversity in the selected areas of the lake resulting in higher
yields of fish and improved fish catches for the fisherfolk. It was hoped
that the improved fish yield would convince the fishermen of the success of such scientific techniques of restoration.

To encourage the participation of the local fisherfolk in restoration
efforts the project promoted the use of cheap, locally available materials for the scientific experiments so that it would be easy for the local
communities to replicate the successful experiments of restoration.
Thus for enhancing algae growth, ropes that were locally made were
used. For the creation of brush parks that act as fish aggregation

devices, palm fronds were used and locally available tiles were used for batteries, encouraging colonisation by juvenile prawns and crabs.

The project also sought the involvement of the marginalised Yannadi tribal community in setting up experiments of ecological restoration. Yannadi women who handpicked juvenile prawns were to be trained in raising prawn hatcheries for stocking indicator species. The project team planned to elicit the cooperation of the local fisherfolk in the imposition of a year-long ban on the capture and sale of berried crabs and of juveniles of tiger prawn and flower prawn.

3. Pulicat Lake and its ecology

Pulicat Lake, the second largest brackish water lagoon (after Chilika Lake in Orissa) in India, has an area of approximately 461 sq. km. Situated between 13°24′ and 13°47′ latitude, and 80°2′ and 80°16′E longitude, this remarkable lagoon harbours huge fisheries sprawling across the two coastal states of Andhra Pradesh and Tamil Nadu. The lagoon is about 60 km in length and varies between 0.25 and 17.5 km in breadth. It is extremely shallow, because of silt deposits (Mathew 1991). The lake is confluent with the Bay of Bengal with a sand bar of 3 km stretching north from Pulicat Lighthouse on the Tamil Nadu side of the lake.

This spectacular lagoon is an important refuge for migratory and resident waterfowl. Endangered migratory species including the spot billed pelican and flamingo visit. The famous Buckingham Canal constructed for famine relief during British rule runs through the lake. The canal was also used for inland trade between the villages around the lake. The lake was declared a bird sanctuary in 1976 although it took 14 years for the Collector to settle the rights of those living in the area in accordance with the procedure specified in the Wildlife (Protection) Act. The final notification of the area as a sanctuary was issued on 30 May 1990. The lake had an impressive total waterfowl count of 83,806 in 1988, but in 1992 the total waterfowl count had plummeted to a mere 10,800 (Rao and Mohapatra 1993). The lagoon also harbours several species of prawn, crab and molluscs. According to Dr Sanjeeva Raj, the principal scientific investigator of the project, the numbers of tiger prawn, flower prawn, mud crab, cockle, clam and green mussel which were once abundant in the lake have drastically dwindled in recent years. The biological diversity of the lake today includes 160 species of fish (Kaliyamurthy 1972), 25 species of polychaetes (Sunder and Raj 1987), 12 species of penaeid prawn, 29 species of crabs

(Joel and Raj 1988) and 19 species of molluscs (Thangavelu and Raj 1988). Fishing has been and remains the main occupation of the 35,000 people living in the 52 villages around the lake. Pulicat Lake has been a major fishing centre from ancient times and was an important trading post for the Portuguese and Dutch in the sixteenth and seventeenth centuries. The few records of the Department of Fisheries indicate that in 1971–72 the total production from the lagoon was 800 metric tonnes, of which about 50 per cent comprised prawns (mainly *Penaeus Indicus*) and about 20 per cent mullets (*Mugil Cephalus*). Of the total landings from the lagoon, over 60 per cent are landed in Pulicat town alone.

The best fishing sites are found on the southern Tamil Nadu side where the salinity of the lake is relatively less and evaporation is less extreme. In the summer months, excessive siltation closes the mouth of the lake. The fishermen living in the traditional *padu* villages, especially those near the lake mouth, manually open the mouth of the lake in summer in order to maintain the brackish water quality of the lake.

4. The *padu* system

Since the project envisages the participation of the local fisherfolk in the conservation of the lake, it is necessary to describe here the fascinating institution of the *padu* that governs the life of the traditional lakeside fishermen living in the area. The fishermen of Pulicat Lake follow a unique system of fishing called the *padu*. The *padu* system specifies access rights to the fishing grounds, and was originally practised by the lakeside fishing villages of Nadur-Mada *kuppam* (also called Christian *kuppam*), Kottai *kuppam* and Andi *kuppam* on the Tamil Nadu side. *Padu* draws the lakeside fisherfolk into a system of sharing or equitable access to the fishing grounds, and this traditional system governs fishing practices, regulates fishing and other allied social activities amongst the traditional lakeside fishermen belonging to the *pattanavar* caste. Today the fishermen of Pulicat Lake are a heterogeneous group with the dominant traditional lakeside fisherfolk practising the *padu* system of fishing. There are also maritime fisherfolk who go to sea to fish and non-traditional fishermen consisting of tribal groups of Yannadi and Irulas, and the predominantly Muslim fisherfolk in the boat-building village of Jamilabad who do not fish in the traditional fishing grounds but in the peripheral regions of the lake. In the last two decades, the traditional *padu* system has been extended to include other villages located on the sand bar.

Principles of equitable access to the fisheries of the lake and collective responsibility of conservation of these fisheries were woven into this traditional system of fishing. As the system applied to all the traditional fishermen and regulated fishing in southern region of Pulicat Lake, it also functioned as a traditional conflict-resolution mechanism. *Padu* literally means 'fishing site' and refers to a system of rotating access to a fishing ground (Mathew 1991). The fishing system has been reported to be prevalent in some south Indian and Sri Lankan fishing communities as well. The traditional fishermen of Pulicat Lake assert that the *padu* is an age-old system that has been practised by generations of fishermen living around the lake. The *padu* system can best be described as a traditional system of granting entitlements to eligible members of a particular community to undertake specific fishing activities in certain designated fishing grounds of the lagoon during specified seasons (Mathew 1991). Remarkably, this system prescribes every minute detail involved in the fishing operation and even specifies the permissible fishing gear and the boats that can be used.

The *padu* system applies only to the traditional fishermen of Pulicat Lake who belong to the dominant fishing caste called the *pattanavar*. (*Pattanavar* literally means 'one who lives in a town' or *pattanam* (Mathew 1991).) The fishermen of the *pattanavar* caste generally live at the southern end of the lake. They are full-time fishermen because the southern part of lake does not completely dry up in the summer. (The fishermen living towards the north of the lake are only seasonal.) Married *pattanavar* youth of 15 years of age and above are eligible to become members of the *talekattu*, the village-level organisation of fishermen. The fisherman seeking membership of the *talekattu* should be skilled and acceptable to the village community. As a member of the *talekattu*, he is expected to shoulder responsibility for common village expenditure such as litigation, temple repairs and festival expenses.

The fishing grounds within this system are divided into three zones, the *Vadakku padu*, the *Munthurai padu* and the *Odai padu*, with the most productive fishing ground being the *Vadakku padu* and the least productive the *Odai padu*. An astute observation made by Mathew (1991) is that although this system of fishing operates on the principle of equitable access to the fishing grounds, it does not ensure or even attempt to ensure equitable returns from the fishing grounds of the lake. Thus, different fishing gears bring in varying yields.

There are about nine distinct types of fishing gear used in and around Pulicat but the two most important are the shore drag nets and the stake nets which are referred to in Tamil as *suthu valai* and *badi valai*.

(The Tamil word for net is *valai*.) *Suthu valai* consists of two components, the *tadukku* and the *siru valai* (Mathew 1991). The *tadukku* is a wall-like net about 2.4 m in breadth, which is fixed with casuarina poles. The other component, the *siru valai*, is a big bag-like net (broader at the bottom and narrower at the mouth) and is about 13 m long. The lower side of the net is secured to stakes driven at the bottom of the lake while the upper end has floats to keep it open. Fishing is mostly done at night, during the low tide, when the prawns migrate to sea. The *tadukku* functions as a wall and stops the movement of fast-moving prawn. When the prawns encounter the barricade, they swim against the current and are entangled in the *siru valai*.

The *badi valai*, which was introduced in Pulicat around 1905, is a symbol of power and affluence. It is essentially a drag net almost in the shape of shore-seine, mainly used for catching mullets and other species, especially during the neutral phase of the tide. The operation of *badi valai* involves 25–30 people and about three to four *padagu* (plank boats). The entire operation requires about 80 people. There is a division of labour in the use of *badi valai*: some undertake the actual fishing operation, some prepare the gear for the operation and others dry the net after the operation.

The *badi valai* is owned by a privileged few (village heads) whereas the *suthu valai* is common to all fishermen. The *badi valai* is not species-specific, whereas the *suthu valai* is especially designed to catch prawns. The *badi valai* is now only occasionally used in Pulicat owing to the alteration of the fishing grounds after a massive cyclone that hit Pulicat in 1984. Consequently, many of the fishing grounds that were ideal for the use of *badi valai* have become uneven making it impossible to use this net.

The *padu* system for the *suthu valai* operates on a lottery system for the eligible *talekattu* of the villages. Each village, independently of the others, carries out the fishing operation. Every *padu* village knows from the beginning of the year the days designated for the village for fishing in the specified fishing ground. On days considered auspicious, all the *talekattu* of the village meet to draw lots to allocate the fishing grounds allotted to the village. Interestingly, different *padu* villages use alternately the most productive fishing ground (*Vadakku padu*) as well as the less productive fishing ground (*Munthurai padu*) thereby ensuring equitable access for all the fishermen of the villages. According to the fishermen belonging to the *padu*, only 14 days in the month are considered good for fishing, they are: full moon day, new moon day and three days each before and after the full moon and the new moon.

During these 14 days there will be heightened tidal activity enabling the active movement of prawns towards the sea (Mathew 1991). Each *padu* village has, on average, five days in a month allotted for fishing. The varying productivity of the fishing grounds further restricts the number of fishing days; fishermen often forsake their access rights to the less productive *Munthurai padu* and wait their turn in the *Vadakku padu*. In effect, each village has access to the best fishing grounds once in 30 days, although the entitlement to fish is for 120 days in a year. The fishermen state that the day the lots are drawn, they can predict the catch potential in their allotted fishing grounds in that particular year: the potential catch of prawns of different fishing sites is common knowledge for the fishermen.

Although the *padu* system embodies the principle of equitable access to the fishing grounds of the lake, it is clear that equitable access is for fishermen belonging to the *pattanavar* caste only (even Christian fishermen of Nadur Mada kuppam belong to the *pattanavar* caste). The 'new' fishermen are not permitted to use the main fishing areas where the *padu* system operates and have been allowed to operate only on the periphery of the traditional fishing sites. The Irula and the Yannadi, who were tribal food-gatherers and marginal cultivators respectively, have been allowed to handpick crabs and prawns during the day in the southern end of the lake. Both Yannadi men and women stand in neck-deep waters in the lake for hours hunting for crabs and prawns with their bare hands. Sometimes the Yannadi use simple small rafts and simple fishing gear, called *katcha* to hunt for crabs. The Yannadi and the Irula survive on small-scale subsistence fishing and are amongst the poorest in the area. The 'new' fishermen are strictly prohibited the use of plank boats or fishing nets in the lake.

5. Evaluation of the project

It is surprising that a project that regards the participation of fishermen as an integral component in the endeavour of ecological restoration of the lake has totally neglected or overlooked the institution of *padu*. This is a serious shortcoming because the *padu* not only organises fishing activities in the lake but also sets the pattern of the wider social life in the area. One reason for the indifference shown by the traditional fishermen to the restoration work conducted by the scientific team is that few attempts were made by the funding agencies or the concerned partner organisation to mobilise the *padu* to communicate the objectives of the project to the local fishermen. The scientific team did not get adequate feedback on their experiments and efforts of conservation because they unwittingly bypassed the *padu* system and could not communicate the usefulness of these experiments to the fisherfolk. As there was no sociologist or anthropologist working with the scientists (from CRiNIEO), the scientific team could not fully appreciate the central role that *padu* could play in fulfilling the objectives of the project. For instance, as reported earlier, the scientific team wanted the local fishermen to ban the fishing and sale of berried crabs, and juveniles of tiger prawn and flower prawn, for a year in order to stabilise the populations of prawn and crab. The *padu* is a strong institution which could by consensus support any proposed activity of restoration, and could offer wholehearted support for any measures banning the fishing of juvenile prawn or berried crabs. The importance of the *padu* struck me when I worked jointly with the fishermen to lobby government authorities about the disastrous impact of development activities and mega-development projects on the fishermen and the ecology of Pulicat Lake. To my surprise, I found that the *padu* fishermen of Pulicat could within minutes mobilise leaders of nearly all the villages in and around Pulicat Lake for a meeting to decide on the course of action they should collectively adopt in connection with the construction of the port.

Despite being a strong and effective institution for regulating fishing in Pulicat Lake, the *padu* system has obvious shortcomings, especially as it excludes the 'new' fisherfolk: the poor Yannadi, Irula and the relatively better off Muslims of Jamilabad. The 'new' fisherfolk have taken to fishing in the lake in the last decade or so in order to survive and augment their livelihood. So far, they have not questioned the traditional institution that excludes their participation on an equal basis, although a few minor skirmishes between the *padu* and the new fishermen have been reported. As the pressure of population dependent on the lake increases, the *padu* system may disintegrate, leading to reckless exploitation of the lake. To avoid such an outcome, it would be necessary to enable the fishermen of the lake to preserve the *padu* system and extend its principles to accommodate the 'new' fisherfolk. Without such an extension or modification of the institution, any attempt to promote the participation of the fishermen in conservation will remain partial at the best. For instance, how can the Yannadi, who barely survive by handpicking crabs and prawns, comply with the proposed ban on the hunting of berried crabs and juvenile prawns? If this has to be done, the Yannadi should be allowed access to some of the traditional fishing grounds. The *padu* has in the past been extended to some newcomers of the *pattanavar* caste from other villages who were not originally conferred rights to fish in the lake.

From the point of view of project management, a major hurdle to the success of the project has been the cynicism displayed by the traditional fishermen who play a dominant and vital role in decision-making. There is a widespread feeling among both the lakeside and maritime fisherfolk that restoration efforts of the kind envisaged by the project team cannot reverse the massive disturbance and alteration of the ecology of the lake that would be caused by construction of the port. They believe that the lake ecosystem is seriously imperilled by the port under construction at Ennore, which is a mere 20 km away from Gunankuppam at the southern tip of the lake. The port is being built to supply and transport coal to the North Madras thermal power plant, located at Ennore. According to the fishermen of the area, the port would accelerate erosion of the already badly eroded sand bar separating the lake from the Bay of Bengal – a view widely endorsed by renowned scientists and ocean engineers. The fisherfolk say that the operations at the power plant have already affected the quality of the water in the lake, and marine pollution has drastically reduced their catches. They reason that the projection of the new port into the sea will divert tidal pressure to the sand bar, thereby eroding it even more. They expect tidal pressure to breach the sand bar at a few points and convert the brackish water lagoon into a salt-water lake. They anticipate depletion of brackish water species such as prawns and mud crabs and are worried that their catch of prawn and crab may collapse. As they do not possess the fishing gear or the knowledge and skills required to venture into the sea, there is a widespread fear amongst them that they may soon lose their only source of livelihood and face an uncertain and bleak future. If this major ecological change does takes place, it would destroy the *padu*, which is not only the traditional system of cooperative fisheries management, but also a way of ordering the social life of the fisherfolk. Possible erosion of the sand bar threatens the maritime fishermen as well. They fear that their homes on the sand bar will be washed away by the increased tidal pressure of the sea, thus forcing them to migrate. They are perturbed at the prospect of displacement because they have already suffered the trauma of relocation when they were uprooted from their homes on Sriharikota island further north, to make way for the establishment of the Indian Space Research Organisation. They also fear the depletion of fisheries in the sea owing to increasing sea traffic and pollution generated by the port at Ennore.

It is clear that the original log-frame of the project did not anticipate or take note of the possible ecological harm caused by development projects initiated by the government on Pulicat Lake. On numerous occasions the fishermen have bluntly asked the project staff about the rationale behind their work. How could we as the project team even attempt to restore the lake and its endangered species when the existence of this brackish water lagoon is clearly threatened by the development of the port?

To answer such questions honestly, the project management staff of the eco-restoration project attempted to get information from the port authorities and the District Collector about the possible impact of the port on the lake ecology and fishermen of Pulicat. Our investigation revealed that the authorities responsible for the port project had conducted no systematic environment impact assessment of the port. Further, they had also bypassed the requirement of mandatory public hearings where objections of the affected persons could be heard and taken into consideration.

The lakeside and maritime fisherfolk at Pulicat have contemplated legal action to challenge the highly questionable approval given by the Ministry of Surface Transport to the port project at Ennore without adequate planning and measures to safeguard their livelihoods. They have asked the project team for assistance through legal intervention and publicity. It must be noted here that the legal intervention and the submission of an independent rapid environment impact assessment of the port sought by the fisherfolk would be a new objective for the project. As a project coordinator, it was extremely challenging and difficult to address such unanticipated problems. Prior ecological and sociological research in the region could to some extent have anticipated some of the problems that the eco-restoration project encountered. It is now clear that any effort for ecological restoration of Pulicat Lake would have to address the ecological and socio-economic costs of this port. These new changes would necessitate further research, which would be a separate study in itself.

Another grave problem affecting the lake ecology of Pulicat is that the practice of intensive and semi-intensive aquaculture for the export market, which was started on a large scale in the early 1990s in the Pulicat Lake, causes irreversible damage to the quality of the brackish water of the lake. The artificial feed pumped into the aquaculture enclosures is dumped along with the wastes into the lake, leading to widespread pollution and possible depletion of already endangered fish species. The intake of salt water necessary for cultivating prawns also increases the salinity levels in the lake. In addition, the outbreak of viruses in these prawn farms has adversely affected the fishing grounds of the lake, along with depriving the fisherfolk of their traditional access rights to fish in the lake. In fact, the Supreme Court of India, in a landmark judgment (*S. Jaganath* vs. *Union of India*), banned intensive and semi-intensive aquaculture within the Coastal Regulation Zone (CRZ) area and declared them illegal. The judgment gave specific orders for Pulicat and Chilika Lakes and stipulated that no aqua-farm would be allowed within 1,000 m of these lakes. Further, the Court held that the existing aquaculture farms, which have been operating illegally within the CRZ, would be liable to compensate affected persons and fishermen. The Court also ordered that the workers employed in the shrimp culture industries, which are to be closed under the order, shall be assumed, for legal purposes, to have been dismissed with effect from 30 April 1997. In spite of this historical order, aquaculture farms, owned by people with strong political connections, continue to operate on the lake. The owners even obtained a temporary stay on the judgment.

It is important to note here that although the fishermen of Pulicat are protesting against the adverse ecological effects of mega-development projects initiated and approved by the government, they are not totally averse to taking up employment at the port. They are acutely concerned about their livelihood and are not primarily concerned about the ecological degradation of the lake *per se*. Clearly, ecological considerations are secondary to their livelihood interests. From time to time they have demanded that they should be given jobs in the port or in the government elsewhere as compensation for the loss of their only source of livelihood. As a matter of fact, the thermal plant at Ennore has absorbed a few of the fisherfolk of Pulicat, thereby raising their hopes that the others may also be accommodated if sufficient pressure is brought on the government. It is telling that the decision-makers and concerned authorities in the government have not even considered the question of the livelihood and culture of the fishermen while hastily granting approval to development projects such as the port at Ennore.

6. Conclusion

In hindsight, if projects such as the eco-restoration project at Pulicat are to succeed in involving local people in conservation efforts, it is important to formulate the objectives of the project, after first carrying out a thorough sociological and ecological study of the area and its people. It is also important to address directly the livelihood needs of the local people and weave them into the objectives of conservation, in order to continue the work of restoration and sustain their participation in conservation activities envisaged in the project.

Further, there should be sufficient flexibility in the project framework in order to address the impact of unanticipated problems that may hinder the objectives of the project. In the case of the Pulicat project, for instance, it should have been possible for the project team to extend support to the fishermen's campaign against the construction of the Ennore port and facilitate the fishermen's campaign by conducting and funding an independent environmental impact assessment study. NGOs such as the WWF will need to modify their project objectives to include proper scientific investigation of the ecological impact of the port, and funding for legal intervention to demand the right to information for the affected people. NGOs working for conservation of threatened ecosystems should realise that it is vital to understand the symbiotic relationship between the fishing communities and the lake ecosystem. In other words, for conservation to be effective, it would have to be directly linked to the livelihood needs of the people.

Note

1. The chapter is based on my experience as the project manager of the project on *Eco-restoration of Pulicat Lake with Fisher-folk Participation* undertaken by the WWF-India. At the time I was working as a Programme Officer in the Wetlands Division of WWF-India.

5

Land Husbandry for Sustainable Agricultural Development in a Subsistence Farming Area of Malawi: Farmer Adoption of Introduced Techniques

Max Kelly

1. Introduction

In this chapter I discuss the contribution of an externally funded development project (promoting soil conservation) to the sustainability of local agricultural systems among smallholder farmers in Malawi. Soil erosion was first recognised as a problem in Africa in the early 1920s. In British-ruled territories (including Malawi) soil conservation became a major issue in the 1930s when a number of schemes were started (Anderson 1984). Despite considerable efforts with soil conservation projects during and since colonial times, however, most have failed (Stocking 1985; Reij *et al.* 1986; Hudson 1987; Pretty and Shah 1994). Evaluation of soil conservation projects in sub-Saharan Africa has shown that little long-term success has been achieved (IFAD 1986; Food and Agriculture Organisation 1991). Maintaining the soil base is a prerequisite for sustainable agricultural production, which has been recognised since the colonial interventions. Until recently the promotion of soil conservation, as well as other areas of development, has been based on the 'transfer of technology'. This top-down approach involves solving problems perceived by the 'foreign' agencies involved in development projects (Hudson 1995). Technologies are often introduced from other areas or countries, but are then frequently found to be unsuited to local climatic or socio-economic conditions. The top-down implementation of these techniques involves decision-making by project managers and organisations, often with little or no input or feedback by the supposed beneficiaries.

The failures of this conventional development paradigm have led to a demand for alternative approaches to development projects (Chambers 1983; Chambers *et al.* 1989; Hudson 1995). This is especially important in subsistence farming communities. In many cases these farmers have few resources and little access to credit, drastically reducing their ability to use fertilisers, pesticides or improved seed varieties for their crops. Also priorities of subsistence farmers can be very different, leading to risk-averse farming strategies rather than increased production. Soil conservation is a long-term strategy and can often have no short-term benefit, leaving it as a low priority for a farmer struggling to produce enough food every year to feed his or her family (Food and Agriculture Organisation 1991; Pretty 1995).

The basis of the new paradigm is a holistic model of development that incorporates social, cultural and environmental concerns as well as economic factors. Central to this paradigm: the participation of the farmers in the development process; acknowledging local knowledge and local institutions; and using a 'bottom-up' approach (Chambers *et al.* 1989; Pretty and Shah 1994). In terms of soil erosion Hudson sums this up as a change of attitude from soil conservation to land husbandry. Land husbandry incorporates better farming, which includes increased productivity and water management, as well as soil conservation. Land husbandry also involves preventative as well as curative soil conservation measures (Hudson 1995).

Here I examine these new approaches with particular reference to soil conservation and land husbandry, using research on the Promotion of Soil Conservation and Rural Production project (PROSCARP) in Malawi. I use the adoption of techniques introduced by the project as a base to explore where the project is succeeding or failing, looking at the farming communities and the agricultural systems in operation. Specific examples of adoption are presented to ascertain success rates achieved within the farming population currently trying to implement the project techniques. In conclusion I discuss the changing approaches to development typified by the PROSCARP project and assess the reactions to this by the farmers in both their opinions and actions.

2. Introduction to the study area

Malawi, a land-locked country situated in the southern part of the East African Rift Valley in southern Africa, has a land area of 118,500 sq. km, of which 80 per cent is land and the remainder is water, dominated by Lake Malawi. Lake Malawi is the southernmost of the Great Eastern Rift

Valley lakes and the rift valley escarpment runs through the country, which has a topographical range of 37 to 3,050 m above sea level. Lake Malawi is 474 m above sea level (United Nations and the Government of Malawi 1993).

Malawi has highly variable temperature, rainfall and vegetation. The Malawian climate is tropical, with three seasons. A cool, dry period, May to August, is followed by hot weather (very hot in low-lying regions) during which humidity builds up until the rains commence in November/December. The rains peak around the turn of the year and continue intermittently until April. Rainfall varies countrywide from 600 mm to 3,000 mm. Most of the country receives sufficient moisture for rain-fed agriculture, although rainfall is becoming more erratic in recent years and droughts (such as that of 1991–92) are becoming more common (Bunderson *et al.* 1995).

The last census, in 1988, estimated Malawi's population at 8.2 million. The population is presently growing annually at a rate of 3.7 per cent. Population density is lower in the Northern Region and higher in the Southern Region: in the Central Region, where this study is based, it is estimated at 113 persons per km^2 (United Nations and the Government of Malawi 1993), one of the highest population densities in Africa. Malawi is also one of the world's poorest countries and has few mineral resources. About 85 per cent of the population work in the smallholder farming sector. Smallholders raise cassava, maize, pulses, cotton, tobacco, groundnuts, rice, fish and livestock (Ministry of Agriculture 1995b). Malawi is a former British colony and achieved independence in 1964. A one-party state, led by Dr Hastings Kamuzu Banda ruled until 1994, when democratic elections were held. The country is now ruled by the United Democratic Front (UDF).

2.1 Soil erosion in Malawi

Land degradation is one of the major problems in Malawi (Ministry of Agriculture 1995b). Traditional farming practices were extensive, involving shifting cultivation. Population pressures in the central region of Malawi have forced continuous cultivation that does not allow fallowing. The intensification of agriculture has added to the country's susceptibility to erosion because of the topography, rainfall pattern and the erodability of the soils. Cultivation of marginal land is also becoming widespread. Soil loss was estimated to average 20 tons per hectare per annum with rates as high as 50 tons per hectare per annum in some areas (World Bank 1992). Good land husbandry practices can reduce this figure to as little 0.15 tons per hectare per annum (Amphlett 1983).

Soil conservation in Malawi dates back to the beginning of the twentieth century when the European presence was consolidated. Construction of physical soil conservation measures began in the late 1920s with recommended measures including storm drains, silt pits, terraces, contour stone walls, contour hedges and use of cover crops. From 1945 a government conservation campaign was implemented with over 118,000 km of earthen bunds constructed. From 1955 mechanical construction of dams and terracing on steep slopes was introduced (Ngoleka n.d.). In 1968 (under the newly independent government) the World Bank-financed Lilongwe Land Development Programme constructed 288,000 km of bunds, 2,573 km of crest roads, 7,725 km of diversion ditches and 933 km of artificial waterways using government machinery (Ngoleka n.d.). All these measures were to address the problem of runoff.

The initial promotion of conservation was carried out through a country-wide education programme. This proved to be unsuccessful and a 1946 law provided for fines or imprisonment for farmers not following prescribed land use methods. Conservation was turned into a political issue by African Nationalists and was used to attack colonial administrators. After independence the legislation was moderated and promotion of conservation measures was again carried out by education or incentives or a combination of both (Ngoleka n.d.).

3. Introduction to the case study villages

In all the villages the chief controls the land. The land is held in customary tenure with the boundaries marked out by the traditional authority (TA). Security of land tenure does not appear to be of concern within the villages, despite the fact that in all villages the number of farm families has increased substantially over time. The chief allocates land as more households are set up. The amount of arable land left for allocation is very small in all the villages, but none of the chiefs appeared concerned by this.

Table 5.1 gives basic descriptive information about the five villages involved in this research. Villages in italics are those not covered by the PROSCARP project. Mbatamila catchment is situated on the lakeshore plain and has a maximum of 15 per cent slope. The majority of land is within the 0–5 per cent slopes with higher degree of slope being found along the hills of the forest reserve bordering the village or along the banks of the Ngodzi river running through the village.

Table 5.1 Research sites within Salima ADD

Catchment	Village	Project involvement	Number of farm families	Elevation[1] (metres above sea level)	Average land slope (%)
Mbatamila	Mbatamila	Yes – since 1989	385	500–35	0–15
	Chifuwa	No	400	487–518	0–10
Naluva	Chigoneka 1	Yes – since 1996	96	594–670	0–25
	Chigoneka 2	Yes – since 1996	75	594–670	0–20
	Sanga	No	250	570–630	0–20

[1] Lake Malawi is 474 m above mean sea level.

Mbatamila and Chifuwa villages have been there as long as anybody can remember.

Naluva catchment is situated at the base of the rift valley escarpment and shows slopes of up to 25 per cent. Chigoneka 1 and 2 were established in 1974. People in the village were displaced from the Lilongwe area when the new capital city was built. Previously they were farmers on the flat fertile plains surrounding Lilongwe. They grew the same crops, but Chigoneka is much more sloped and has less fertile land. The government compensation was insufficient to build a new house or prepare the land for farming.

Arable farming on slopes of over 12 per cent without appropriate interventions to prevent soil loss is contrary to conventional land husbandry regulations (Hudson 1995). However, shortage of more suitable land, especially in Naluva catchment, leads farmers to cultivate very steep slopes.

Devereux (1997) found that landholding size was a reliable indicator of poverty and food insecurity in Malawi. The Government of Malawi asserts that small smallholders (<0.5 ha) are unable to produce sufficient food and would require 'targeted income transfers' to protect their food security in the short term. Medium smallholders (0.5–1.0 ha) would have the potential to achieve self-sufficiency if their agricultural productivity is improved (Ministry of Agriculture 1995a). Table 5.2 shows the average landholdings and the proportion of households surveyed who fall into the above categories.

Table 5.2   Average land holdings within the study area and percentage of small-holders farming less than 1 hectare

Catchment	Village	Average land holding (ha)	Households with less than 0.5 ha (%)	Households with 0.5–1.0 ha (%)
Mbatamila	Mbatamila	1.2	7.5	30.2
	Chifuwa	1.0	15.7	33.3
Naluva	Chigoneka 1	0.9	10.3	53.9
	Chigoneka 2	1.0	6.3	40.7
	Sanga	0.8	27.3	43.7

Source: Questionnaire survey.

4. The PROSCARP project

This research is based on a European Union (EU)-funded development project. The project is Promotion of Soil Conservation and Rural Production (PROSCARP), and it started in 1989, when it was known as the ADD 'food for work' programme (ADDFOOD). It began as a food and inputs for work programme, providing a basic package of food maize, improved hybrid maize seed and fertiliser to the farmers. The main problem was seen, at that time, to be the need for smallholder farmers to leave the land for cash employment (weeding, planting, and so on for larger farmers). This 'Ganyu' work takes the farmers away from their own land at the time they need to till, plant and weed. They are unable to produce enough from their own land to see them through the season without seeking Ganyu work, and the project tried to break this cycle.

The emphasis of the project has changed considerably since 1989, to concentrate on soil conservation, agroforestry and public health. The aim now is to 'develop strategies for use by resource poor farmers to tackle problems of poor food security caused by declining soil fertility' (Ministry of Agriculture 1995a). The project goals were stated as:

- the wide-scale implementation of agricultural extension strategies for the promotion of smallholder farming systems that can sustain food production;
- increased use and profitability, overall, of hybrid maize, hybrid sorghum and fertiliser through the adoption of alley cropping associated with soil and water conservation;

- proven technical strategies and packages for increased food production and security;
- improved nutrition and public health (Ministry of Agriculture 1993).

Overall the main strategies are:

- pegging and construction of marker ridges along the contour;
- alignment of ridges to contours;
- boxing and tying of ridges;
- construction of waterways and diversion channels;
- filling minor gullies and checkdams;
- alley cropping (for soil stabilisation and as a source of biomass);
- field tree establishment of *Faidherbia albida*;
- nutrition education, shallow well construction and sanitation platforms for pit latrines, to improve health and therefore labour availability;

Strategies differ between sites slightly, as in tree species used for agroforestry. The following techniques are currently being implemented on a trial basis:

- the use of Vetiver grass (*Vetivera zizanoides*) on marker ridges and on the edges of gullies to stabilise the soil;
- minimum tillage in 0.2 ha plots (one plot per village) to investigate the potential of a technique that has been quite successful in Zimbabwe.

Although the project no longer uses incentives for work, PROSCARP still provides significant inputs in the form of seeds, both improved maize varieties and alternative foods such as soya, cowpeas and pigeon peas. Agroforestry species are supplied either to plant out, as in Vetiver grass, or for nursery establishment. Crop seeds supplied are on a revolving fund basis, to be repaid with interest after harvest.

PROSCARP operates nation-wide in Malawi, but it does not attempt to cover every smallholder. Catchments within each Agricultural Development Division (ADD) are selected for project intervention. At the end of 1996 there were 172 sites divided among all the ADDs. The number of sites is planned to expand to over 1,500 by the year 2001 (Ministry of Agriculture, 1995b). Sites are known as catchments – areas within which the project is working, not a hydrological watershed. Each catchment can consist of one or several villages and vary immensely in size. The research was carried out within the Salima RDP (Rural Development Project). The 12 catchments in Salima RDP are all separate and quite distant from each other. The number of catchments grows each year, although Salima ADD is not planning any more expansion until other ADDs are more fully covered.

The project has finished its first five-year period and has secured funding for a further five years from the EU. Efforts have been made to phase out the complete project control in catchments already in operation. Catchment committees are expected to be able to take over the day-to-day running of the work in progress in each catchment. This would include the training of the farmers in marking of contours, planting and maintenance of agroforestry species, as well as running nurseries. The aim is to have each catchment self-sufficient within three years.

The Ministry of Agriculture (MoA) promotes the use of contour planting and improved crop varieties as well as the use of agroforestry species. The Field Assistant (FA) for each catchment is the extension agent for villages in the PROSCARP project and other villages within the Extension Planning Area (EPA). Due to lack of resources from the MoA, however, few inputs are available for these villages, which rely instead on advice from the FA about contour planting and crop varieties to improve their farming methods.

The basic aims of the PROSCARP project are in line with the current government's policies. Since the early colonial period food self-sufficiency has been a strong element of government policy (Devereax 1997). The current government is focused on the overall goal of the reduction of rural poverty, one of the means of achieving this is seen to be the improvement of food self-sufficiency (Ministry of Agriculture 1995b).

5. Participatory methods within the PROSCARP project

In Pretty's (1995) typology of participation, scaled as a way of interpreting the term participation, passive participation is the lowest form of participation, and self-mobilisation – the greatest – is the goal towards which participation in development should aim (see Table 1.1 above, p. 3). At the outset of the project in 1989 under the title ADDFOOD the type of participation would fall into the category of passive participation or participation for material incentives. This type of participation is low on the scale. As the project has matured it has changed. The PROSCARP project has moved into the category of

functional participation. This can be seen in how the catchment committees were set up. The function of the catchment committees is set
out in the next section; the Ministry of Agriculture hopes that by pass-
ing responsibility to the farmers they will become self-sufficient. The
project has initiated Beneficiary Assessment Surveys, carried out in
1993, 1994 and 1996 (Ministry of Agriculture 1993; Leach and
Marsland 1994; Leach and Kamangira 1996). These surveys, carried out
using PRA, show a tendency towards interactive participation. The surveys were, however, carried out in a limited number of catchments and
were used more as a survey for the purpose of assessment than as dedi-
cated move to allow villages to take further part in the planning and
implementation of the project. The results of the PRA exercises were
brought to project headquarters. It is unclear whether any of the recommendations contained in these reports has been acted upon. The
project staff showed a willingness to encompass the concepts of farmer
participation into the project and the project has changed considerably
since its inception.

Thus a movement is occurring towards a more participatory approach,
whilst not yet approaching any form of self-mobilisation. The results
presented in this chapter are used to examine the different approaches
used by the project and the difficulties in changing the approach from
an incentive-led approach to a more participatory approach.

5.1 Catchment Area Development Committee (CADC)

One of the base concepts in participatory development is the use of
community groups or committees. Within PROSCARP each catchment area has a Catchment Area Development Committee (CADC).
PROSCARP provides assistance in soil and water conservation, soil fertility improving measures, sanitation and water. This process is supposed to allow the farmers to practise and adopt 'sustainable forms of
resource conserving technologies' (Ministry of Agriculture 1995b,
Annex 7: 4) After 2–3 years the farmers should be independent of regular extension advice in key areas. The CADC will become responsible
for organising and coordinating the various activities the villagers have
chosen to implement. Existing community groups are contacted for
inclusion in the project work. Where no such group exists, the villagers
will be required to form a committee. The CADC is essentially the link
between the community and the project staff. They are responsible for
drawing up and revising a Catchment Area Development Plan (CADP),
for communicating and assisting cooperating farmers, the community

and the project staff. The CADC should be composed of at least 40 per cent female members (Ministry of Agriculture 1995b).

Since the number of staff and time available in the nation-wide expansion of PROSCARP are limited, the CADC must play a vital role in helping the farmers within a CA become self-sufficient and effective at the project technologies. If extension agents are required to put in further time and effort in one of the CAs, then expansion to other sites is not possible or as effective. Focus group discussions with the members of the catchment committees in each of the project villages did not give a very optimistic picture. All the committee members felt that they passed information between the extension agent and the other members of the village, but, none of them felt they had any responsibility for running the project or any influence in determining project activities in the catchment area development plan. None of them felt able to take over any activities from the extension agent, such as information dissemination, training in project techniques or supporting roles. The committee members agreed that powers of decision-making and change were still firmly in the hands of PROSCARP staff.

Following on from the power relationships in participatory development is the purpose of participation. Participation can be seen as either a means or an end in the development process – a theme that comes through strongly in recent literature (Lane 1995; Nelson and Wright 1995; Chambers 1997). Participation can be used to achieve pre-set goals by a funding authority or implementing agency. In this instance, participatory development or popular participation is used to increase beneficiary awareness and adoption: here participation is a means within the process. If a community or group sets up a process to control its own development, this would be participation as an end. In both cases there is a strong difference in the power relationships within, as well as external to, the community. PROSCARP is externally funded to achieve pre-set goals and the incorporation of participation into this project is as a means to achieve these goals. This raises difficulties in achieving self-mobilisation, as a prerequisite would be the goals of the project being strongly in line with the priorities and needs of the farmers.

6. Constraints on production for smallholder farmers

6.1 Farmer ranking of problems

In each of the villages a ranking exercise was carried out to ascertain the major problems or priorities of the communities. Table 5.3 shows

Table 5.3 Ranking of problems in each village by the village members (1 being the most important issue)

Problems	Mbatamila	*Chifuwa*	*Chigoneka* 1	*Chigoneka* 2	*Sanga*
Food scarcity	4	3	1	1	1
Soil fertility	5		2		
Soil erosion	7				
Disease	1	2	4		7
Pests	6				4
Lack of cash for inputs (seeds, fertiliser)	3	5	3		5
Water supply	2	1		3	2
Access to markets		4		4	6
Access to hospital		4		2	3

Source: Questionnaire survey.

the results of the exercise. In each village only the problems that were suggested by the villagers were included. Food scarcity was ranked as the most serious problem by all three villages in the Naluva Catchment, whilst disease and water supply ranked as the two most serious in Mbatamila catchment. All the villages suggested that the remedy to most of these problems would be the ability to earn cash income, but they could see no way to increase their current ability to do so except through increasing the amount of cash crops, such as cotton or tobacco. Land scarcity means that increasing the area of crops grown is not practical without reducing food crops. Increasing the yield of cash (and food) crops is reliant on the use of fertilisers and pesticides, which are very expensive.

This exercise allows some comparisons to be drawn between the priorities and needs of the farming communities involved in the project and the goals and objectives of the project itself. The goals of PROSCARP concern problems such as low food production, inadequate water supply, poor soil fertility and disease. Soil erosion, however, is a very low priority in Mbatamila Village and was not ranked at all in the other villages. Soil fertility as a limiting factor in food production was an important priority for the farmers, but cash to purchase inputs such as fertiliser was of greater importance. The basic conflict between PROSCARP goals and the priorities of the farming community appear to be based around that of time-scale. Soil conservation is a long-term strategy in terms of reducing erosion. Increasing soil fertility through the use of biomass from agroforestry species is, depending on the establishment time for the species used, a medium-term strategy. Village priorities are a reliable food supply, health and water. These are all short-term priorities.

7. Adoption of PROSCARP techniques

The intervention of the PROSCARP project can be divided into the following areas:

- soil conservation and crop production;
- health and nutrition;
- sanitation;
- water.

The reasons for adoption or non-adoption of techniques for each of these areas emerge from a survey of households showing how many actively participate in the project in each of the three project villages (see Table 5.4). Mbatamila has been involved in the project since its inception in the 1989–90 growing season. The other two villages have been involved only since 1995–96 growing season.

Mbatamila has the highest proportion of households involved in the project. Since the project started more people have become involved over time. Table 5.4 shows that even among villages that are involved in the project, many households have not implemented project techniques. The following sections examine household uptake of project strategies.

Table 5.4   The total number of households who feel they actively participate in the PROSCARP project

Name	Participate	Do not participate	Total number of household in the village
Mbatamila	48 (91 per cent)	5 (9 per cent)	53
Chifuwa	0	51 (100 per cent)	51
Chigoneka 1	28 (70 per cent)	12 (30 per cent)	40
Chigoneka 2	27 (84 per cent)	5 (16 per cent)	32
Sanga	0	54 (100 per cent)	54
Total			230

Source: Questionnaire survey.

7.1 Soil conservation and crop production

7.1.1 *Physical conservation measures*

There are two main strategies promoted by the PROSCARP project. The first is the promotion of contour planting, including the construction and planting of marker ridges (every sixth ridge) with the ridges in between realigned to the marker ridges. The second is the use of agroforestry species both for soil conservation and for soil fertility. The number and total amount of marker ridges constructed is shown in Table 5.5.

Mbatamila Village has been involved in the project for the longest time and has the highest percentage of households participating in the project (Table 5.4). However, a much lower proportion of households within this village has a high proportion of marker ridges constructed on their land. Chigoneka 1 and 2, although recent entrants to the project, both have nearly twice as many households who have completed marker ridge construction. When marker ridge construction was discussed at the village meetings, two explanations were offered for the successful construction of marker ridges.

First, farmers feel they need help from the Field Assistant (FA) to mark out the ridges using the A-frame, a simple tool to mark contour lines where the marker ridges are constructed. Nobody in the meetings said that they were confident using the A-frame without the FA present. The limited amount of time an FA can spend in each village severely reduces the number of marker ridges pegged.

Second, many farmers saw no need for the construction of marker ridges. In Mbatamila, which is situated on less hilly ground, the majority of farmers who were in favour of the construction of marker ridges were those who had land either adjacent to the hills at the back of the village or along the river banks where there is more slope. In some cases farmers opposed contour planting as it can lead to waterlogging of the crops.

Despite the problems of labour availability in female-headed households there was no significant difference ($p < 0.05$) between gender of the household head and project participation or between gender of household head and the percentage of land with marker ridges planted. This topic was discussed with female groups. The marking and construction of marker ridges is carried out during the dry season, when labour requirements for agricultural work are at their lowest, allowing more time to complete the work. The major labour shortages felt by the female-headed households are during the growing season when crop production must take place as well as daily chores, and when off-farm employment can be found in larger estates.

Table 5.5 Numbers of marker ridges constructed by village

Village	Percentage of households in each category				
	No marker ridges	1 to 33 per cent of land with marker ridges	34 to 66 per cent of land with marker ridges	67 to 99 per cent of land with marker ridges	100 per cent of land with marker ridges
Mbatamila	30.2	13.2	17.0	3.8	35.9
Chifuwa	80.4	2.0	3.9	0	13.7
Chigoneka 1	12.8	0	18.0	2.6	66.7
Chigoneka 2	18.8	0	12.5	0	68.8
Sanga	70.9	7.3	9.1	0	12.7

Source: Questionnaire survey.

7.1.2 Agroforestry

The main agroforestry species promoted by the project has been *Leucaena leucocephala*, a fast-growing, drought-resistant, nitrogen-fixing tree, which can be used for green manure, soil conservation and live-stock feed (Young 1989; Bunderson *et al.* 1995). *Leucaena* is used for alley cropping, planted on the marker ridges (at 4.5–5.4 m apart depending on the degree of slope), to improve soil fertility both directly through nitrogen fixation and indirectly through pruning and application of biomass as a green manure. *Leucaena* also helps stabilise the marker ridges (Bunderson *et al.* 1995). *Leucaena* has been used in Mbatamila since the project began in 1989. Over 66 per cent of marker ridges planted in Mbatamila use *Leucaena* (40 per cent and 45 per cent respectively for Chigoneka 1 and 2). Although the farmers present at the meeting were aware of the benefits of growing *Leucaena* they expressed serious reservations about the species, citing problems of seed shortages, establishment and survival. The use of biomass from *Leucaena* as an alternative to artificial fertiliser was seen as a good alternative. However, none of the farmers present had managed to harvest sufficient biomass to achieve noticeable improvements in crop yield.

Other agroforestry species promoted are *Sesbania sesban*, *Glyricidia sepium*, *Senna spectablis* and *Tephrosia vogelii*. All these are leguminous, contributing to soil fertility. Pigeon pea (*Cajunas cajun*) is also used as a soil-improving crop and is planted by some farmers on marker ridges.

In Mbatamila farmers were happy to plant *Glyricidia sepium* and *Senna spectablis* for green manure and soil improvement, but also encountered difficulties with establishment and yield. Overall in Mbatamila the main drawback to the use of agroforestry species as an alternative to artificial fertiliser is the problems of establishment. The yield of the current species is also far too low to be a realistic fertiliser alternative. However, as cash or credit to purchase artificial fertilisers was not available to most of the farmers present, there seemed to be an overall willingness to keep trying.

Ascertaining how successful agroforestry species are in Chigoneka 1 and 2 is more difficult, as these villages have been active in the project only since 1995–96. This was a year of very low rainfall followed by a year that had abnormally high rainfall. This has presented difficulties in tree establishment and for those that have established agroforestry species on their land it is too soon to show any real benefit. The farmers were enthusiastic about the potential benefits, as the majority of them are unable to obtain credit for the purchase of artificial fertilisers.

Vetiver grass (*Vetiveria zizanioides*) has been introduced in the last two growing seasons as a drought-resistant, deep-rooted, perennial grass species, mainly used for soil stabilisation on marker ridges, steep slopes and gullies. Farmers are currently being paid by the PROSCARP project to plant Vetiver nurseries, making it difficult to ascertain farmer opinion of this strategy.

Farmers are also provided with seeds for soya, cowpeas, pigeon peas and, in some cases, improved maize varieties. These are distributed on a revolving fund basis but do not appear to be paid back by many of the villagers. The impression from the farmers was that the seeds were given free.

The one species that was unanimously recognised as being of great benefit in all villages is *Faidherbia albida*, locally known as Msangu. This tree is native to Malawi (Bunderson *et al.* 1995) and occurs in every village. This tree loses nutrient-rich leaves during the rainy season so improving soil fertility whilst allowing crop growth under the canopy. Yields of maize crops under the canopy can show increases of 50–250 per cent compared to crops outside the canopy area (Bunderson *et al.* 1995) and increases from 50 to 100 per cent in soil organic matter and nitrogen have been reported under the canopy (Young 1989). As the tree species is already present in the villages, if in limited numbers, the benefits are easily seen in the healthy vigorous growth of maize under the canopy. The PROSCARP project is seeking to establish systematic intercropping of Msangu at 10×10 m intervals. Other benefits of Msangu include its use for wood fuel and building materials. Some of the problems with Msangu are the long-term return on investment as tree maturation is 7–10 years. Also care is needed in planting out as the tap root is sensitive. The majority of farmers appeared enthusiastic about this tree species, with the reservation that artificial fertiliser would be needed to bridge the gap between tree planting and tree maturation. Many farmers already had one or more Msangu trees growing on their land and had kept them because of their soil-improving capabilities.

7.1.3　Firewood

Wood is the main source of fuel in the villages. Wood for fires is no longer available in the village and must be collected from further away. Women gather the wood once or twice a week. The average distances travelled to the site of wood collection are set out in Table 5.6.

Wood fuel used to be available in the villages. The women said that each year they have to travel further to collect sufficient wood for

Table 5.6 Average distance walked for the collection of firewood (in km)

Village	Average distance walked for firewood	Maximum distance walked for firewood
Mbatamila	1.0	1.5
Chifuwa	1.3	4.0
Chigoneka 1	2.4	9.0
Chigoneka 2	1.1	4.0
Sanga	1.6	7.0

Source: Questionnaire survey.

cooking. Wood is now collected from uncultivated hillsides or protected forest reserves. The PROSCARP project does not address the problem of wood fuel supply. Agroforestry species used are for soil stabilisation and fertility, and although some firewood may be produced it is minimal.

7.2 Health and nutrition

The Women's Home Assistant carries out visits to the project villages. The programme includes education for the women about household hygiene, rubbish disposal, food preparation, nutrition and childcare. Discussions were held in each of the project villages with the women's group. They were enthusiastic about this aspect of the project as they could see the benefits in reduced diseases and improved child health from implementing the information gained in the education pro-gramme. Although the health and nutrition programme was well received it has actually caused problems in the overall opinion of PROSCARP, as visits from the Women's Home Assistant have now become infrequent. In Chigoneka I and II no visits at all were made during 1996, leading to a lot of discontent as the women are now very aware of the benefits of the programme since it has started but have been denied the chance to learn more.

7.3 Sanitation

PROSCARP also promotes the use of pit latrines and the provision of sanitation platforms (sanplats) to cover the pit latrines. PROSCARP encourages households to dig a pit latrine and has provided materials for the construction of sanplats. Table 5.7 shows the percentage of households who have dug a pit latrine and those who have a covered pit latrine. The villages in the PROSCARP project show no noticeable difference in the total number of pit latrines from the non-project villages. However, the number of sanplats installed is much higher in

Table 5.7   Percentage of covered and uncovered pit latrines in the villages

Village	Households with uncovered pit latrine (%)	Households with pit latrine and sanplat (%)	Total houses with sanitation (%)
Mbatamila	15.1	35.8	50.9
Chifuwa	60.8	5.9	66.7
Chigoneka 1	10.3	38.5	48.7
Chigoneka 2	18.8	53.1	71.9
Sanga	52.7	0.0	52.7

Source: Questionnaire survey.

the project villages due to the requirement for materials to make the sanplats.

The benefits of sanitation were well understood. In Mbatamila it was mentioned that the pit latrines and sanplats were sufficient at the beginning. However, due to the increasing number of households and people it is now necessary to construct more.

7.4   Water

The availability of a clean, protected water source is an important safeguard against disease. The women collect the water twice a day. The distance to the water source and the amount of time waiting to collect water can severely impede women in terms of time available for other chores.

In Mbatamila in 1992 four shallow wells were dug in the village with the help of the PROSCARP programme. They are 4–5 m deep. Usually one or two of them run dry. At the time they were dug they provided sufficient water, but as the number of households has increased so too has the demand for water, and queues at the wells can often exceed one and a half hours. Some people now use the river again, which is not a safe source of water.

In Chigoneka 1 there are two shallow wells located in the village. The first was dug in 1995 and the second in 1996. Both were supplied with pumps but one of the pumps is now broken. PROSCARP was responsible for the boreholes. Both wells run dry by the end of September or October and water must be fetched from the river. Chigoneka 2 has a shallow well dug by the villagers near the river in 1994. The water department provided a pump (now broken) and water is lifted by means of tins attached to strings. The well runs dry by July or August and needs to be dug deeper. This requires external help, as the ground is too hard to deepen the well with traditional hand tools. During the time that the well is dry the villagers dig shallow wells along the riverbank for water.

In all the project villages the main benefit of PROSCARP was seen as the provision of shallow wells, despite the fact that the wells do not provide water all year around and are not always sufficient for the number of people in the village.

7.5 Farmer opinion of PROSCARP

A survey (see Table 5.8) was carried out to determine farmer opinion of the various aspects of the PROSCARP project. The survey asked whether PROSCARP contributed to the farmer's ability to:

1. feed his or her family;
2. improve the fertility of the soil;
3. improve the health of his or her family;
4. improve access to safe water.

Table 5.8 shows that the most positive response to PROSCARP is related to the water programme, and this also came across very strongly in the village meetings. Although the wells can run dry during the dry season, the availability of clean water in the villages for most of the season is a considerable improvement in quality of life in the villages. This also reduces the incidence of water-borne diseases and in many cases reduces the total amount of time that the women must spend on household chores.

The ability of the PROSCARP project to improve health in the family is also recognised. This is in response to both the education received about health and hygiene, as well as the availability of clean water and

Table 5.8   Results of a survey of farmer opinion towards the PROSCARP project

Village	Food (per cent positive)	Soil fertility (per cent positive)	Health (per cent positive)	Water (per cent positive)
Mbatamila	50.9	43.4	41.5	75.5
Chigoneka 1	23.1	20.5	43.6	66.7
Chigoneka 2	31.3	40.6	62.5	75.0

Source: Questionnaire survey.

sanitation facilities. The least positive responses were about the ability of the PROSCARP project to increase the fertility of the soil and to help increase food production. Table 5.8 shows that food scarcity was ranked as the most serious problem in three of the five villages. Agroforestry is the main focus of the PROSCARP project with the long-term aim of both increasing soil fertility, and thus food production, and reducing soil erosion. The lower opinion of this aspect of the project is explained by the problems related to agroforestry. These are set out in detail in the previous section on agroforestry.

8. Discussion

The PROSCARP project has achieved successes, with many people involved in the project. The water, health and sanitation components of the project are quite successful. This acts as a good basis as it encourages positive reactions from the beneficiaries to the project overall. However, it appears important that these aspects of the project do not degrade over time. The failure of the Women's Home Assistant to visit both Chigoneka 1 and 2 in the previous year lets the whole project down.

Techniques such as contour planting and alley cropping are being implemented in the villages. The rate of uptake is different between the older site, Mbatamila, and the more recent project entrants, Chigoneka 1 and 2. The farmers involved in the project were aware of the potential benefits of agroforestry. The main problems seem to be related to the management and yield of the species planted on the marker ridges. The majority of farmers have realised few if any benefits from the alley cropping.

The PROSCARP project is attempting to adopt a more participatory approach, but the potential benefits of this have yet to be felt. The catchment committees are supposed to be strongly involved in the day-to-day implementation of project techniques, but these committees show no obvious signs of taking on more responsibility or becoming more self-reliant. Techniques such as pegging of marker ridges using an A-frame are still dependent on the field assistant helping out. The potential for farmers to participate is strongest here. The opportunities provided by the catchment committees are numerous. Theoretically, the Catchment Committees should become self-sufficient and act as a base for the village to manage its own development. This requires the empowerment of the Catchment Committee to decide the future of the project, to select what fits their needs and to reject what is not suitable.

This illustrates one of the paradoxes of the use of participation as a means to achieve pre-set goals. Although the objectives and goals of PROSCARP have changed considerably since the start of the project in 1989, the changes were not initiated by the beneficiaries themselves. The project receives funding to promote soil conservation. Although the villagers are aware of the problems of soil erosion they are much more concerned with soil fertility. Contour planting and the use of Vetiver grass on marker ridges are both specifically to control erosion. Alley cropping and interplanting of soil-improving species can do much to increase the fertility of the soil, but it is here that the most problems are felt.

PROSCARP is reliant on the participation of the farmers to carry on the project cycle while project staff can decrease support and move on to other sites. At the current time this is not happening. There is a strong argument here for a much greater change in project management and implementation. It would require time, effort and training to increase the participation of the farmers, assuring that each of the project villages is self-sufficient in both knowledge and inputs and that the project reflects the needs and priorities of the beneficiaries. This would lead to a greater chance of the benefits of the project being felt after the withdrawal of project support.

Part III

Learning from Success: Where National Policies are Supportive, but Participative Action is Novel

6
Local Management of Sahelian Forests
Paul Kerkhof

1. Introduction

According to project reports, Sahelian forests have been the subject of participatory management since the 1970s.[1] What has continuously changed is the definition of participation and of community. The contribution of local labour to reforestation was an important criterion of participation in the 1970s, and in particular the participation of women. Large tree-planting schemes were established, often in reaction to perceived or real desertification in peri-urban areas or in sensitive areas such as borehole perimeters.

Villagers' willingness to establish wood lots on their own land with cost-sharing arrangements was considered the right form of participation from the early 1980s, and it was implemented in a large number of village forestry projects in the Sahel. Elsewhere in Africa this strategy was also pursued, particularly in countries with a socialist orientation, such as Tanzania and Ethiopia (Kerkhof 1990: 216).

The tree species as well as the tree planting and management techniques were essentially the same as those used for large-scale plantations of the 1970s, with the same foresters in charge of the operations. The added participatory elements were that village land was used instead of gazetted land, unpaid village labour was encouraged, and management decisions were in principle made by the village government. In practice, the forest service remained a key decision-maker, for legislative reasons.

The village wood lot concept declined in the late 1980s, when it was found that the majority of wood lots grew poorly, they were hardly managed at all, and villagers' main interest in the wood lots was merely a response to the insistence of intrusive government officials.

Various studies noted the vast financial cost of reforestation in the drylands on any significant scale. Growing seedlings, planting and protecting them are essentially very hard jobs in arid and semi-arid areas, and yield poor returns. Clearly, local people's participation was unlikely to improve much.

2. Strategic shift to natural forest management

The Guesselbodi project, Niger, and TRDP in Turkana, Kenya, were among the first to involve villagers in managing the remaining natural forests, instead of trying to restore them where they no longer existed. Locally meaningful forests management units were identified. In the case of the Sahel, these usually involved neighbouring villages, each of which was apportioned a nearby section of the forest. In Guesselbodi, management plans based on forest inventories were drawn up, and a contract was signed between the village committee and the forest service on the basis of the management plan. The forest service remained the key player.

In Niger, this strategy was increasingly seen as the proper strategy of making local communities responsible for the local forest resources. A World Bank/Danish-funded project, Energie II, prepared fiscal policy changes to reward local forest management. Woodfuel from organised village markets was taxed less than wood fuel from 'non-organised markets' elsewhere. The project also improved the understanding of the Sahelian forest ecology: based on wood fuel harvesting trials, it recommended simple forest conservation techniques. The final objective of Energie II concerned the sustainable supply of wood fuel to the urban markets (Foley *et al.* 1997). The Rural Woodfuel Market model developed in Niger (1989–96) was replicated in Mali in 1996, and is currently being prepared for Senegal, Burkina Faso and other countries with assistance from the World Bank and northern donors.

Parallel to the Rural Woodfuel Market, the 'Gestion de Terroir' approach has been developed, based on land use planning at village level. Unlike the Woodfuel Market it takes a multisectoral approach to forest management. Gestion de Terroir is an integrated land use planning strategy in which commercial firewood production is but one of many aspects. The Gestion de Terroir and the Rural Woodfuel Market approaches are similar in that both identify the village as the basic management unit to which responsibility should be transferred, and through which external assistance should be channelled. A constraint in the Gestion de Terroir approach with respect to forest resources is that forestry legislation, especially in the francophone Sahelian countries, puts the villager in a very weak position vis-à-vis the forester. The contracts and the fiscal regime provided under the Rural Woodfuel Market model are intended to put the villager in a stronger legal and commercial position.

Research and development workers have expressed concerns about the Gestion de Terroir and the Rural Woodfuel Market approaches, which both identify the village as the land use planning and management unit. Both approaches assume that the village is a community without internal differences, but stakes in the forest resource are varied and interests are therefore often contradictory. Because natural resources tenure institutions in the village are often dominated by descendants of the first cultivators, that is to say, the oldest lineages, more recently established migrants may be poorly represented, even if they constitute the majority of the inhabitants. Furthermore, nomadic land users, who have important and long-established stakes in the forest, are often under-represented, or not represented at all. Nomadic land use patterns cross village boundaries and sometimes national frontiers (Anon. 1992; Paris 1997). In the following sections, I explore both internal differences in the farming communities and herder–farmer conflicts at some depth.

3. Villages as multi-stakeholder communities

The Rural Woodfuel Market objective is sustainable woodfuel production for urban markets. Ecological research, technical innovations, fiscal policies and legislative changes are shaped to suit this objective in the countries that accept the system (Kerkhof 1990; Peltier 1994; Foley *et al.* 1997). A criticism that can be levelled against the Rural Woodfuel Market is that many Sahelian forests provide other, more important economic functions than commercial firewood.

Livestock production is particularly important in many forests. Sahelian forests are generally open and constitute an important resource for grazers and browsers. With an increasing proportion of land taken into cultivation, these forests become critical as a grazing reserve during the cropping season. Whilst the economic value of grass and forage resources is notoriously difficult to assess, several studies do compare commercial woodfuel production and livestock sales from the same forest:

- The Kelka forest in Mali is an exceptionally important source of urban firewood. Nevertheless, an economic study shows that 1995

revenue from local livestock is 28 million FCFA (without incorporation of nomadic livestock) while revenue from firewood sales is
19 million FCFA. Firewood revenues are dropping at about 20 per
cent per annum due to rapid depletion of deadwood (Anon. 1997b).

- Calculations in Baba Raffi forest, Niger, indicate that annual revenue
  from livestock equals 340 per cent of the annual revenue from firewood sales (Paris 1997).

- An ongoing study in the forest of Tyi, Mali, shows that the 1997 revenue from livestock is in the order of 4 million FCFA, against an
  annual commercial firewood production well below one million
  FCFA (Anon. 1997b; Kerkhof 1998).

Furthermore, the artisanal and construction wood produced in many
forests may be more valuable than firewood. Studies of the Tyi forest in
Mali and El Ain in Kordofan show that the revenue from artisanal and
construction wood sales is much more important than the revenue
from firewood sales. A study of artisanal and construction wood production in the flood plain forests of Zinder, Niger, arrived at similar
conclusions (J. Madougou, personal communication 1996). Finally, in
sample villages of the SOS Sahel study fruit production from natural
forests commercialised by village women is more important as a source
of revenue than firewood.

None of the Rural Woodfuel Market-oriented studies indicate that
woodfuel is more important than other economic values of the same
forests. Elaborate economic monitoring systems, limited to the urban
firewood trade, have been put in place, but little is known about the
impact of this market system on other sectors. Given the very high
funding level of the Rural Woodfuel Market system, it is also peculiar
that none of the studies has assessed the distribution of benefits over
the various local stakeholders or social groups in the village.

The SOS Sahel studies show that benefits may be socially or spatially
concentrated. Forest exploitation is hard work, which is done in the
absence of better alternatives. In one study area, the village owning
the forest is not interested in commercial wood production, since it
has a dry season water sources for commercial gardening. The neighbouring village does not have this option and most families depend
on forest exploitation for an income during the dry season. For the
poor families who make up over half of the total, the forest is the
major source of revenue, which enables them to pay their taxes and
fill their cereal deficit. The poorest families, however, hardly benefit
from the forest since they do not have a physically able household

member available for this kind of work. Forest users who are well organised and who are in possession of donkey carts are more efficient producers.

In the Malian example, exclusive wood-cutting rights claimed by the village owning the land would have dramatic effects on a neighbouring village. This argues in favour of supra-village level management of such forests. In the Kordofan villages studied, the distribution of benefits is not spatially, but socially determined. Poorer families, though not the poorest, use the forest for commercial exploitation and sell to the richer families in the same village and at local markets.

The use of the forest for grazing tends to be less regulated than wood exploitation in all cases studied. Most families have some livestock but professional herders are obviously the key interest group. In the Rural Woodfuel Market system, problems arise if forest management is entirely dominated by the firewood cutters, strengthened by the forest service. A Nigerian researcher speaks of a socially explosive situation in the study village as a result of resource appropriation by the woodcutters (Madougou 1999a).

A SOS Sahel project in Kordofan has supported villages to register forestland under provisions of the 1989 Forest Act in order to provide long-term security to the forest resource. While this has been successful in most villages, it has run into problems in others. The typical problem is one where the descendants of the first settlers, seen to be the traditional owners of the land (*Dar*, 'home'), have allowed immigrants to settle and cultivate in an allocated piece of land (*Hakura*, 'place for guests'). The immigrants' village was seen as an independent village by the project and by government agencies, but not by the traditional owners. Now that the project is being phased out, some of the traditional owners are reclaiming the forest registered in the name of the immigrants' village (Kerkhof 1999).

4. Herder–farmer conflicts

A growing number of land use conflicts have been reported in the Sahel, in particular in the 1990s. Most of these conflicts appear to concern farmer–farmer confrontations: an important secondary source of conflict seems to be herder–farmer confrontations. A major study in Niger found that 26–46 per cent of families in sample areas are faced with land disputes, many of which are farmer–farmer disputes, with herder–farmer conflicts as an important second category of litigation (Lund 1995). Many herder–farmer and some herder–herder conflicts are extensively reported due to the tragic nature of the conflict, in some cases with dozens or even hundreds of deaths.[2]

Persistent confrontation may turn into inter-ethnic conflict. It is quite possible that such land use conflict is the source of the numerous cases of civil unrest and international conflict in the Sahel, the 'Green War'. The Club du Sahel/OECD suggested that governments and donors should invest more in prevention of conflict and less in military training and equipment (Roy Stacey, 1997, personal communication). Gestion de Terroir, Rural Woodfuel Markets and most other natural resource management projects may be criticised for turning natural resources with complex and contradictory stakes into simplified natural resource management units to the advantage of settled interests whilst excluding nomadic peoples, leading to marginalisation of nomads and possibly contributing to increased civil unrest on the long term.

Although most development agencies have largely ignored the multi-stakeholder nature of the Sahelian forest resource, some organisations are now more careful to study the nomadic interests before supporting a forest management system. Ensuring that all local stakeholders share responsibility for the remaining natural resources seems to be gaining ground.

In a Nigerian SOS Sahel project (Zinder Department), the project ignored nomadic interest in the Takieta gazetted forest for about five years. In 1997 it made a radical move to seek the nomads' participation in Takieta forest management, and in 1998 the project brought the nomadic presence in strategic meetings at par with those of the settled groups. The nomads were very satisfied, saying this was the first time any agency had involved them in land management issues. At the same time, the nomads noted that their interest in the forest is periodic and that their involvement in forest management should be less intensive than those of villagers surrounding the forest. The nomads are now represented in the general assembly but not in committees responsible for follow-up.

In the Samori forest in Mali, an SOS Sahel project found it hard to support herder interests. One constraint in the project, and in most forest management projects for that matter, is that all project staff are from a settled origin with little appreciation of the nomadic culture and economy. In 1998 the Peuhl leadership identified two Peuhl herders to work with the project on issues affecting the herders. They function as 'local researchers' who have had formal education and who understand the culture of settled peoples, but who are still active herders. This project is now part of the SOS Sahel (UK) and IIED

Drylands Programme pastoralist research programme supporting eight Sahelian natural resource management projects.

5. Constraints of multi-stakeholder projects

Projects that aim at the involvement of the various stakeholders and the negotiation of their interests tend to have less specific objectives. The logical framework, which has become the standard project planning tool, has a final objective, which is normally something vague like 'improved livelihoods' or 'higher socio-economic standards'. Participatory forest management is then accepted as an intermediate objective and not an end in itself, and firewood production is just one of various concerns. NGO projects are usually more experimental, participatory and more holistic than government projects, but the cost of scaling up may be prohibitive. This distinction between the more sectoral government projects and more multi-sectoral NGO projects is not unique to Sahelian forest management (Farrington 1998).

The more holistic approach taken by many NGO projects has led to an extensive programme of activities, the 'mésures d'accompagnement' which may well overshadow the original sustainable forest management objective. There is an urge to address short-term problems before dealing with longer-term environmental issues. Gender divisions have often been recognised and special support to women has been provided by many projects, especially through women's credit schemes. Many other needs – such as health, water, transport, adult education, institutional development – are also addressed. The critical issue in many of these projects is impact evaluation when new funding is sought at the end of a phase. Evaluators and donors alike wonder how much impact the project has had in terms of sustainable, participatory environmental management – the main objective for which funding was provided in the first place.

Impact assessment at the long-term objective level in the logical framework is complicated in many sectors. Danida's Agricultural and Natural Resources Sector Review pointed out that it is virtually impossible to measure project impact in isolation of other factors in development such as national and international economic trends, rainfall, or political change (DANIDA 1993). In the natural resources sector, still dominated by the subsistence economy, impact measurement is probably more complicated than in other sectors.

Forest monitoring is traditionally done through inventories, which measure cubic metres of wood, based on European forestry tools. The amount of fieldwork required and the statistical complications are a heavy financial and human resources burden. Consultants invariably carry out such inventories. Project staff rarely grasp the concepts and consultancy reports are hardly used, while the inventories are completely beyond the local communities who are the ultimate forest managers.

SOS Sahel has developed two alternative tools that can be mastered by local people. Local product inventories count the number of locally defined forest products such as building poles by local producers, so that the results can be directly related to local economics and to rule-making. Still more simple is panoramic photography, by which villagers take panoramic photos (360 degrees) of their forest at known places, collate the photos, and keep them in the village for future reference. The latter technique does not require any literacy but still attains a degree of objectivity. The limit to any codified form of forest monitoring, though, is that they carry some cost, whereas the economics of Sahelian forest management outside a project environment hardly justifies such cost. The alternative to codified monitoring is what local people have always done, which is traditional, subjective monitoring at no cost: observe how the forest is changing and communicate verbally.

Measuring the impact of community, project and policy interventions in the Sahelian forest resource is a key weakness. The methodological difficulties are enormous. Monitoring physical and socio-economic parameters in project and non-project areas may be required over a long period of time, but it is never done. Yet the stakes are high. Donors may not continue to invest in participatory forest management if impact cannot be proven. The best proxy to long-term impact monitoring is probably analysis of case studies.

6. Cases of local forest management

Four case studies of local Sahelian forest management are presented here to illustrate some of the impacts and some problems of impact measurement.

- In the 5th Region of Mali, the Near East Foundation has assisted 15 villages since 1991 in securing exclusive local rights on collection and sales of firewood and other forest products, the most important source of income for the villagers. Most resource conflicts between villages, and between villages and town-based traders, are now resolved by a local supra-village institution. Interference in forest exploitation by the forest service through harvesting permits to outsiders is minor. Combined revenue (1992–96) from forest products is more important than that of agriculture or livestock, though declining since stocks of deadwood are rapidly diminishing. While the project invested considerably in forest inventories and economic monitoring, sustainable exploitation quotas remain guesswork (Sylla 1996; Anon. 1997b). The main impacts are improved internal conflict management, strongly reduced exploitation by town based traders, and reduced dependency on forest service agents.

- In the same Region, SOS Sahel (GB) has supported 50 villages in Bankass to rid themselves of exploitation by traditionally corrupt local forest service agents. SOS Sahel concluded a convention between the forest service and the traditional resource management institutions that are now responsible for rule-making, control and sanction. Differences between the villages and non-project villages, where foresters still impose illegal penalties, and between pre-project and present management have been noted in economic terms. It is clear that some social groups and some villages depend on these forests for economic survival (food purchase and tax payments). However, the ecological sustainability of forests under local management is as yet unknown and the legal framework of the conventions is weak (Kerkhof 1996; Anon. 1997b; Kerkhof 1998).

- Through a World Bank supported project in Niger, 85 Rural Wood-fuel Markets were in operation by 1995 in the Niamey basin, with 84 million FCFA woodcutters' revenue in 1995. Village organisations in principle have exclusive responsibility for forest exploitation and in particular, for sales of firewood, based on a contract with the forest service which stipulates annual quotas. But sustainability is at risk in the absence of continued external support. Powerful urban traders and forest service agents who have lost revenue due to the Woodfuel Markets resist and try to reverse the situation. Other risks are the technical complexity of annual quotas. The social sustainability of the domination by firewood cutters over the village forest is also queried (Foley *et al.* 1997; Giraud 1998; Madougou 1999a). Impact measurement is limited to firewood and related revenues.

- The Woodfuel Market model is now used in Mali by several projects, the oldest being the Kita project which has supported 13 villages by 1997 to obtain exclusive exploitation of state forest as well as village forest resources. Village based woodcutters had a net annual income of 12 million FCFA and a strongly increased share of urban sales by

1996 (Anon. 1997a; BIT 1997). The Malian agency responsible for Woodfuel Market development has noted the technical complexity of the Nigerian model as a serious weakness. Forest inventories in Mali are executed by one private agency (BICOF), which is considered as the only competent agency in forest survey. This is prohibitive for widespread adoption of the Woodfuel Markets. Attempts are made to develop a forest monitoring system which is manageable by village institutions and which does not need well-trained foresters (H. Konandji, 1998, personal communication).

- In northern Kordofan, Sudan, SOS Sahel (GB) has assisted 17 villages to obtain ownership certificates of their natural forests, the only examples existing in Sudan since adoption of the Forest Act 1989, which provides for the legal context. Ecological monitoring in some forests shows that forest regeneration is better inside registered forest than outside, due to protection by villagers against fire. Furthermore, several registered forests have withstood land expropriation by powerful companies that expropriated land under other statutes (Kerkhof 1999). The forest service requires a management plan for each village forest, even though none of the 160 state forests in north Kordofan have a management plan. The most important *de facto* management by villagers is protection against tree-cutting and against fires, of which various indicators are available. But it is evident that nomadic groups increasingly resent what they see as exclusion from their traditional grazing areas. The project recognises that omission of nomadic interests is a key weakness in the strategy.

7. Key issues

When analysing these and various other Sahelian forest management experiences, several key issues become apparent.

7.1   Balance between science and participation

A balance must be found between 'scientifically correct' and participatory forest management. The drive for a sustainable forest management system, as defined by the forestry service, puts a massive strain on the management system. Projects which want to put in place a 'scientific system' require classic forest surveys, mastered by few experts at high cost. Dubious presumptions are made about growth rates in order to arrive at exploitation quotas of various kinds, and the management transaction costs are too high (Anon. 1988; Foley *et al.* 1997; Hopkins 1992; Mounkaila 1997; Sylla 1996).

The ecological characteristics of the Sahel jeopardise the usefulness of the quota system. Periods of good or poor rainfall change the forest ecology rapidly with important consequences for management, whereas the quota system assumes a stable production pattern. Furthermore, villagers adjust to agricultural productivity in a form of economic nomadism. They may be engaged in natural resource exploitation for some years but not at all in others, depending on conditions. A recent World Bank review suggests dropping the quota system that is so characteristic of the Woodfuel Market (Foley *et al.* 1997). In support of the 'laissez-faire' model, some senior World Bank staff noted:

> It seems prudent, as a general policy, to transfer authority permanently, and then let local people experience the consequences of their own management decisions. Such transfers should be preceded by careful discussion with local people on the terms and conditions of the transfer. If they maintain or develop the natural forest, they will derive the benefits; if they reduce or destroy it, they will support the cost ... it is certainly no more risky for natural forest maintenance than continuation of the official control measures which are presently inadequate in most areas. (Foley *et al.* 1997)

Some projects have pursued the alternative 'laissez-faire local management system' such as the Bankass project, or the Norad supported Turkana project, Kenya, of the 1980s (Kerkhof 1990, 1996, 1997a,b, 1998). This essentially leaves the formulation of forest exploitation to local institutions. While the management transaction costs are very low and legitimacy can be high, transparency can be surprisingly poor. In the case of the Tyi forest in Bankass, 14 different rule interpretations exist among 60 respondents. Even members of the local management institution have different rule interpretations. Little is known about the ecological sustainability of such forest management. Sustainability is high in terms of local institutions, but they are not legally embedded.

The indications are that many West African foresters are not prepared to accept significant local management and they wish to maintain quotas, to be determined by foresters or through their support. This is compounded by the rent-seeking behaviour for personal or for institutional reasons. Furthermore, the forest service in the francophone Sahel has the structure of a military service, which is the opposite of what is required for support to local management (Kerkhof 1997a,b, 1998; Madougou 1999). While ex-foresters do some of the most innovative work in local forest management, the traditional forestry institutions in the Sahel show resistance to change. Traditional policing roles still constitute a major source of prestige and income.

7.2 Involvement of nomads

Almost all projects have largely excluded nomadic resource exploitation from their analysis and from the management system. Recent legal innovations in various Sahelian countries facilitate exclusive resource exploitation by settled peoples whilst traditional nomadic rights are poorly represented or exterminated. Examples are the Forest Act 1989 of Sudan (Anon. 1989), recent forest acts in Mali (Anon. 1995a,b) and, to a lesser extent, the Rural Code of Niger (Anon. 1993). Many project reports note that there are 'difficulties with nomads' who do not respect project-supported activities such as rule-making for forest exploitation. Attempt to redress the balance in favour of pastoralists are now made, but will meet difficulties in the face of the current policy and legal climate, and because of poor organisation among nomadic pastoralists.

7.3 Need for new approaches

Even in the 1990s many projects resort to the worn-down micro-catchment and tree-planting concepts when trying to improve forest productivity, yet without signs that it will be any more successful than it was in the 1980s. Some project staff note that tree-planting schemes have been planned or executed in order to encourage forest service officials to accept a project package that includes local forest management.

7.4 Need for legal transparency

The legal arrangements for local forest management in at least a part of the Sahel remain incomplete. A legal study of five Nigerian projects found that none of the supported local institutions had a clear legal status (Sani 1996). Projects make institutional arrangements and ad hoc provisions based on conventions that may be overturned, or contracts that may not be renewed, once the project finance and personnel support is no longer there.

8. Conclusions

8.1 Role of decentralisation

Decentralisation is the key policy change in the Sahel that has encouraged forest management by local communities. Improved participation, democracy and accountability are elements of decentralisation, elements that can be found in improved governance, but hardly in the forest service. The institutional implication is that local forest management should not so much be perceived as a matter of the forest service but of local government. Forms of local government are presently being strengthened throughout the Sahel, albeit it at varying degrees. Alternative institutions, such as environmental agencies, which are multi-sectoral and which do not carry the historic burden of exclusion and rent-seeking, are developing at national level.

8.2 Management issues

A balance between 'scientific forest management' and 'laissez-faire management' has not been established. The former type of management currently dominates due to the historic role of the forest service and due to the forestry legislation. With increased decentralisation and local governance it is likely to shift to the latter type of management. Suitable low-cost tools to strengthen local forest management beyond informal practices have hardly been developed.

8.3 Building capacity at local level

Forest management institutions at local level often lack coherent internal and external rules, accountability, sometimes legitimacy and invariably, legal support. Governance at this level can be reinforced through capacity-building, and with capacities that often exist in the community to manage public services. Apart from better laws, minor external support may be required to achieve this. What is needed most to make it work, is good governance at higher levels in the society.

8.4 Issues of pastoralism

Pastoralists are the principal losers in the examples to date of local forest management. Experience in the Sahel shows that in the long term, impoverishment of this group may lead to considerable civil unrest. Some organisations have begun to involve pastoralists in the local management process, but the experiences are thin on the ground.

Notes

1. This publication is an output from a research project funded by DfID of the United Kingdom. DfID can, however, accept no responsibility for any information or views expressed.
2. See, for example, Lund (1995); Maiga and Diallo (1995); Ba (1996); Konate and Tessougue (1996); Le Roy *et al.* (1996). A conflict between pastoralists and farmers in Darfur, Sudan over the 1998 rainy season has reportedly also led to hundreds of deaths.

7

CAMPFIRE: Tonga Cosmovision and Indigenous Knowledge

Backson Sibanda

1. Introduction

The Communal Areas Management for Indigenous Resources (CAMP-FIRE) is a celebrated approach to natural resource management, developed in Zimbabwe. Now adopted and adapted by many nations in eastern and southern Africa, it has gained widespread popularity among donors, development agencies, academics and practitioners. CAMPFIRE is described in the popular literature as an innovative approach to sustainable management of natural resources which aims at transferring proprietorship, management and utilisation of natural resources to local communities who live in the vicinity of those resources. So far, however, CAMPFIRE has mainly focused on wildlife. The programme recognises that the survival of wildlife and other natural resources ultimately depends on changing the perceptions and attitudes of people who live with these resources. The CAMPFIRE philosophy believes that if ownership, management and utilisation are all transferred to those who live with the resources, these communities will develop positive attitudes towards the resources and use them in a sustainable manner (Murphree 1991).

CAMPFIRE's goal was to put people at the centre of the development process and to let them fully participate in the management of the resources. A major objective was to incorporate local people's indigenous knowledge and traditional practices into the programme. People's knowledge was viewed as providing incentives for communities to own the development process and therefore fully participate. CAMPFIRE recognised that local people had knowledge and experience, which could be utilised in the management of resources. The use of indigenous knowledge and scientific knowledge would create synergies for sustainable management of natural resources.

113

CAMPFIRE was first implemented in valley of the Zambezi River, among the Tonga, who are found on both sides of the Zambezi, in Zimbabwe and Zambia. In Zambia the Tonga are successful farmers and cattle ranchers. In Zimbabwe while they are the third largest ethnic group after the Shona and Ndebele, they are marginalised, poor and with very little education. The Tonga marginalisation started during the 1860s when the Ndebele and Kololo used to raid them for their cattle (Cousins and Reynolds 1993). The lives of the valley Tonga changed forever in 1958 when the Kariba Dam was built and the lake started to form. When the two sluice gates finally stopped the flow of the Zambezi, the Tonga people were forced to move from the valley and were resettled in the escarpment, where the terrain was more ragged, soils were poor and water was scarce (Kenmuir 1978; Tremmel 1996). From then on they had to share their land with wild animals, which hitherto had their own territory.

The Tonga left behind their homes, fields and gardens. They also left behind their ancestral land and their ancestors' graves, and their shrines were flooded by water (Tremmel 1996). Further, they lost contact with their relatives and friends in Zambia, and they left their ancestral spirits in Zambia and could no longer help their relatives there. According to Tremmel, despite great hardships such as illness and high mortality rates, the valley Tonga enjoyed an inner sense of happiness and integrity. He argued that memories of their integrity and happiness continue to overshadow the suffering of their former life. They remember more about their link with neighbours and relatives across the river, the proximity to the water, their fertile fields and gardens and their freedom to hunt and fish. But the Tonga also lost most of their livestock when they moved; their goats, sheep and chickens died during the move. The Tonga were not even assisted to settle in the new areas, but were left to find land for themselves and to build new homes (Tremmel 1996).

The colonial government had ignored the Tonga as a primitive group and left them to live their lives the way they wished until the Kariba Dam was built. Until 1957 the Tonga had continued to manage and utilise their natural resources using their indigenous knowledge. This lifestyle ended when they were moved from the valley and they lost their fishing and hunting rights and their access to fertile land and water. In the valley they had been self-sufficient in food production, but after they moved they became dependent on government food aid – a situation that has continued to this day. When CAMPFIRE was introduced in the Omay-Nyaminyami District it was supposed to redress some of the injustices that the Tonga had suffered. CAMPFIRE aimed to give back the responsibility for management of land, wildlife and other resources to the Tonga and let them benefit directly from them, but above all it aimed to utilise Tonga indigenous knowledge and traditional practices in the management of these resources. CAMPFIRE derived its strength from this commitment to utilise local people's knowledge.

2. The present situation

In Zimbabwe, the government took over the management of resources such as land, forest and wildlife during the colonial period. The government also took the responsibility for conflict-resolution as well as powers of exclusion. The postcolonial administration continued with the same policies. Only recently (since 1988) under CAMPFIRE has policy changed. Past actions have cumulatively had the following results:

1. Traditional institutions such as chiefs and councils of elders have been stripped of their powers (Murphree 1991; Sibanda 1995). These institutions have lost control and influence over natural resources.
2. Competing power centres such as the district council, ward development committees, and village development committees have been established by the postcolonial administration as part of the democratisation process. These institutions now have more power than the traditional ones.
3. At community level, increased stratification has been created by formal education and new local government structures (Birgegard 1993), which may create more problems for the common property resources system.
4. The CAMPFIRE concept aims at restoring proprietorship of resources, including all powers of resource ownership, management, conflict resolution, enforcement or rules and powers of exclusion, to local communities.

If the local institutions do not have powers to enforce agreements, power of exclusion and conflict resolution, the system could easily break down to one of open access, and result in serious resource degradation. On the other hand, common property theories often emphasise the need to be able to exclude some potential users, but such exclusions may pose a new threat unless they are flexible enough to allow for multi-jurisdiction access rights.

3. CAMPFIRE

CAMPFIRE is considered to be an innovative approach to sustainable natural resource management. Maveneke (1992) argues that it amplifies and improves on earlier local natural resources management that placed responsibility for wildlife utilisation and conservation in the hands of local communities. CAMPFIRE's aim is to ensure the sustainable utilisation of natural resources (wildlife, forests, water and soil) in such a way that producer communities (communities who suffer direct inconvenience from wildlife) will benefit directly. These benefits can be in the form of cash incentives to households, community projects, or other benefits such as meat and employment. CAMPFIRE seeks to transfer custodianship of resources in communal areas from the state to the communities by giving 'appropriate authority' or the legal right to manage natural resources to the district councils. The Zimbabwe Trust (1989) described this approach as one which aims at the conversion of 'open access' resources (non-property regimes, or *res nullius*) to common property resources (private property for the group, or *res communis*). It further aims at empowering local people to make decisions and to manage their resources.

The Department of National Parks and Wild Life Management (DNPWLM) developed CAMPFIRE in Zimbabwe. DNPWLM wanted producer communities to be given a full choice of how to spend money generated by their district from wildlife utilisation. This choice included both projects and payments in the form of household dividends. The CAMPFIRE programme, however, gave the local people only the freedom to choose how to spend the income from wildlife, and made no provision for the communities to play any major role in deciding how the resource should be utilised or managed. Significantly (as we shall see later) this creates conflict between the authorities and the local people. The 'appropriate authority' given to the district councils did not give them any power to decide how the resources would be utilised. Custodianship in this case meant limited access. In theory, local communities are supposed to take more responsibility for the management of natural resources, but in reality no legal or practical provisions allow for this to happen.

The genesis of CAMPFIRE came from the realisation that conventional wildlife conservation approaches had failed and that the custodianship of wildlife and other resources must be vested in the hands of local people as it had been in traditional times (Maveneke 1992). In order to transfer this custodianship from the state to the communities some legislative changes were necessary. The 1975 National Parks and

Wildlife Act had given private landowners the right to utilise game for their own benefit. The 1982 amendment gave the district councils 'appropriate authority' or legal authority to manage resources, including wildlife in their areas for the benefit of their citizens. The state, however, retained ownership of land and the resources on it. It is therefore a misconception to view wildlife as a common property resource. At this point however, the state did none the less attempt to address some of the problems of the past by giving local people some access to utilise wildlife and other resources. The weakness of CAMPFIRE has its roots in the amendment to the Wildlife Act, which tried to include other resources that do not fall under the jurisdiction of the Ministry of Environment.

It is probably a misconception to perceive CAMPFIRE as a programme covering the management of indigenous resources (Thomas, S.J. 1991). Thomas also argues that at present the programme focuses on the wildlife resources. The reason for the emphasis on wildlife is that historically CAMPFIRE originated from DMPWLM, which has no authority over resources other than wildlife, and yet wildlife is dependent on land, forests, and grasslands – controlled by different arms of government. While CAMPFIRE has been embraced by other government agencies, only the wildlife component has been implemented. How, then, has CAMPFIRE impacted on wildlife management, and indeed on the management of land and other natural resources in a situation without clear legislation, policies or guidelines?

CAMPFIRE is also caught between what are known as the microcosmic and the macrocosmic views. The microcosmic view interprets the CAMPFIRE concept to mean that only those that carry the costs of living with wildlife should receive benefits from wildlife. This noble position is one that is difficult to implement, given the fugitive nature of wildlife and the resource distribution and management in communal areas. Those advancing the macrocosmic view interpret CAMPFIRE to mean that the benefits from wildlife should be broadly distributed to all communities in the district. This notion too is justifiable. The proponents of the microcosmic view, however, argue that rewarding everyone will kill the incentives for communities to conserve wildlife since even those who have not participated in wildlife protection can receive benefits. These two views have significance for this chapter as they address the management of wildlife under CAMPFIRE. The DNPWLM in 1986 defined the objectives of CAMPFIRE as:

1. To initiate a programme for the long-term development, management and sustainable utilisation of natural resources in communal

areas. The programme would involve forestry, grazing, water and wildlife. The areas covered by CAMPFIRE would be natural regions III, IV and V on the periphery of Zimbabwe. Communities would join the CAMPFIRE programme on a voluntary basis. Natural regions I and II have high agricultural potential given the soil types, rainfall and climatic conditions. Natural regions III, IV and V are dry lands, semi-arid and arid areas with low agricultural potential, but are suitable for ranching and game.

2. To achieve the management of resources by placing custody and responsibility for them on the resident communities.
3. To allow communities to benefit directly from the exploitation of natural resources within the communal areas. The benefits would take the form of income (household dividends) employment and production.
4. To establish the administrative and institutional structures necessary to make the programme work. Administration would be through the established system of Village Development Committees (VIDCOs), Ward Development Committees (WADCOs) and District Development Committees (DDC) to Provincial and National levels.

The CAMPFIRE concept recognises the fact that wildlife survival depends on changing the perceptions of people who live with it. The concept can, therefore, help to restore a perception of wildlife as a valuable resource rather than as a nuisance with no mitigating features. CAMPFIRE is viewed by many writers as creating a powerful incentive for rural people to adopt wildlife management as an adjunct to conventional agriculture (Zimbabwe Trust 1989). The concept is not a blueprint for rural development or wildlife conservation. As a way of thinking, revolutionary in some aspects, it is still evolving. Ecologists, planners, implementing agencies and rural people are changing their perceptions of natural resource economics and conservation.

CAMPFIRE has adopted a multidisciplinary approach, involving wildlife managers, ecologists, sociologists, technicians, economists and legal experts (Hasler 1996). It has also adopted a multi-institutional approach to its implementation by working with government departments, non-governmental organisations, the private sector and with villages, wards and districts (Sibanda 1995, 1996).

The programme is sanctioned by government because of the belief that if people participate in wildlife management and in sharing the benefits that accrue, this will ensure the sustainable utilisation of natural resources (Hasler 1996). Conservation is therefore not seen as an end in itself, but occurring as a result of resource utilisation. According to Hasler (1996) CAMPFIRE makes the following assumptions:

1. involving local people in the economic benefits and management will result in the sustainable utilisation of resources;
2. once people are the proprietors of the resource, they will participate in decision-making and about the management of the resource;
3. the economic benefits targeted for local communities through the district council will reach the local people.

4. Indigenous knowledge

It has been increasingly recognised that local communities, their indigenous knowledge and traditional practices can contribute to the utilisation, management and conservation of natural resources. The potential has been recognised by a few conventional scientists and conservationists, but the majority remain sceptical or have totally rejected the idea that these communities have any knowledge to contribute. Others have treated this as an interesting academic subject, studying how the 'natives behave'. A few groups, out of curiosity, have attempted to implement this on the ground by involving these indigenous peoples in the planning and in incorporating their indigenous knowledge and experiences on resource management. Evidence from CAMPFIRE in Zimbabwe indicates how hard it is for those who are not part of these communities to understand issues of indigenous knowledge, spirituality, taboos, etc. These are still despised, and treated either as superstition or an inferior knowledge system that has nothing or very little to contribute to science and natural resources conservation. Evidence from Zimbabwe and elsewhere in Africa, however, suggests that natural resource utilisation and management are guided by spirituality and a wealth of knowledge accumulated over thousands of years by millions of people (Sibanda 1997).

Some work has also been done to understand the contribution of indigenous knowledge. Many traditional farming techniques – such as indigenous soil and water conservation, ethnoveterinary practices, use of natural pesticides, intercropping and agroforestry – have been assessed and documented. While their importance is increasingly acknowledged, some scientists still dispute whether subsistence management practices of traditional peoples are or were guided by a 'conservation ethic' (Posey 1998). Judgements of the value of the concepts of life, the internal logic of indigenous knowledge, and the process of indigenous learning (cosmovision) and indigenous institutions are highly contentious.

Many indigenous and traditional peoples have sophisticated ways to explain the reality in which they live, and for many indigenous people and rural communities farming is more than just working with the biophysical elements such as seed, soil and water. Farming is a product of the people's cosmovision, which includes the natural world, the social world and the spiritual world. Farming usually involves the application of rules that reflect a notion of a sacred nature, in the belief that a good harvest can only be obtained if farming practices are in harmony with the laws of nature. Community activity is regulated by the rules set by the gods and spiritual beings. For example, amongst the Shona in Zimbabwe, lions are a manifestation of a bountiful harvest, and if a crocodile is returned to the water, this guarantees good rains (Sibanda 1997). The use of natural resources is not exploitation, but a natural responsibility. The dichotomisation of resource use and conservation is a western concept which views resources as things to be exploited. In the Shona cosmovision, human beings are an extension of nature and natural resource utilisation is a celebration of life.

In most African communities natural resources utilisation, management and conservation are a product of the people's spirituality, culture, traditional practices and taboo systems. Conventional scientists do not readily understand these concepts and philosophies, and find it difficult to enter into a genuine dialogue with traditional knowledge systems. Further, confusion has been caused by using and imposing western concepts in interpreting African natural resources management systems. One such concept is the 'wilderness', which clearly demonstrates a very different western cosmovision. Conventional scientists find it hard to deal with issues of spirituality, and have no tools to build the values of indigenous knowledge into management plans. For many African communities the spirits of their ancestors inhabit natural resources. These spirits protect the environment and the utilisation of resources is guided by this spirituality which is directly linked to culture and tradition.

5. Tonga cosmovision and resource utilisation

Conventional science is the dominant paradigm and pacesetter for development in the world today. Indigenous knowledge and people's culture matter only if they fit into this dominant paradigm.

Community participation therefore means people participating in other people's concepts of technology and development. Thus although spirituality is a vital part of the knowledge system of traditional peoples, outsiders do not understand it and do not know how to deal with it. In order to understand how this has affected the Tonga in the last 40 years it is necessary to consider the conflicts between western science and the Tonga cosmovision.

Tonga spirituality, cosmology and the way the Tonga understand the biophysical environment, the elements and how humans relate to physical and spiritual worlds, constitute the Tonga cosmovision. Cosmovision is a concept of life, the perception of an individual or group of the basic principles about the way that the natural (ecological environment), the supernatural (spirit beings) and the human worlds are linked. It includes philosophical and scientific assumptions and ethical positions on the basis of which people relate to each other and shape their practical relationships with nature and the spirit world (Heverkort *et al.* 1996).

In the Tonga cosmovision the natural world, the spiritual world and the human world are on the same continuum, and life is not dichotomised. The spirit is in nature and in humans, hence the natural world is an extension of humanity and the spirit is an extension of humanity (Sibanda 1998). Life is celebrated through the utilisation of natural resources; therefore natural resources are conserved to guarantee the continued celebration of life. There is no resource utilisation without responsibility. For the Tonga, natural resource management and utilisation have a physical, human and spiritual dimension (Sibanda 1998). Life is not seen outside the supernatural because it is the supernatural that controls all life processes. People do not die but move from one world to another, from one form to another. People become shades, ancestors or spirits and watch over the living as well as intercede on behalf of the people to God since the living cannot talk to God directly (Cousins and Reynolds 1993; Sibanda 1998). The Tonga believe that *Leza*, the high God, is the creator and the cause of all that happens. *Leza*, the ultimate power, is not concerned with the everyday affairs of humans. No offerings are made to *Leza* and no priest or prophet claims direct access to the power of *Leza* (Cousins and Reynolds 1993). *Basangu* (spirits that have an effect on the affairs of the general community) speak through people who are possessed. Spirits of animals or foreigners can also cause illness. *Zelo* are ghosts or spirits of dead people who are not ancestors. *Mizimu* are shades or ancestral spirits and they give protection against other spirits (Cousins and Reynolds 1993). *Mizimu* affect only those within the kinship system to which they belonged when they were alive.

The Tonga cosmovision views all natural resources as belonging to God and humans having a responsibility (not a right) of using them in a manner that does not displease God (Sibanda 1998). Natural resources could never be owned, but people have access rights (Cousins and Reynolds 1993). The basic principle of land tenure was the responsibility of a farmer over any land, which he or she brought under cultivation. The community does not own land but oversees it so, for the Tonga, resource utilisation is strongly influenced by spirituality, customs and taboos, and guided by a philosophy of responsible use, and by need not greed. The Tonga believe that a bountiful harvest is not only dependent on the biophysical elements of soil, water and light, but is also dependent on using the resources in a manner that is in harmony with nature and pleases the ancestors and God (Sibanda 1998). Similarly, those who are good hunters are possessed by the hunting spirit to be successful hunters, but the spirit also guides them to avoid killing sacred animals or those that personify spirit beings (Tremmel 1996; Sibanda 1998).

Tonga society is organised in groups called clans, and clans have certain animal names. These animals are not honoured or avoided, but eating the flesh of a clan animal is sacrilegious (Cousins and Reynolds 1993). Each clan protects its clan animals because the vitality and survival of the clan are dependent on the abundance of clan animals. This Tonga culture creates a natural balance in the consumptive use of wild animals. The spiritual and cultural values of the Tonga people were responsible for regulating the use and protection of wildlife. These values are embedded in their cosmology. Tonga indigenous knowledge and traditional practices have often been treated as if they are independent, but they are only a small part of the cosmovision; they cannot be used or implemented outside the broad framework and understanding of their cosmovision. Indigenous knowledge is meaningless outside the people's cosmovision (Sibanda 1998).

6. Tonga indigenous knowledge and CAMPFIRE

On of the objectives of CAMPFIRE is to incorporate local people's indigenous knowledge into the conservation and management of natural resources. The incorporation of local knowledge is part of the people's participation in this programme, because knowledge is one of the most important contributions that local people can make. Through using their own knowledge they can direct their own development. The results of my study, carried out in 1997, reveal that very little Tonga indigenous knowledge has been incorporated in CAMPFIRE. There are several reasons for this failure to meet one of CAMPFIRE's major objectives.

The sample size for this study was derived using the 1992 population census, which provided population figures and numbers of households (Government of Zimbabwe 1992). A 5 per cent sample of all households in the study area was taken, giving a total of 224 households. It was logistically difficult and too expensive to do a larger sample given the population dispersal over a large area.

The household survey was not the only method used in collecting data: in addition 20 structured interviewees and seven oral histories were collected, along with participant observation. This combination of quantitative and qualitative methods was planned to provide an integrated data set which would reinforce each other and provide a rounded picture of the research subject. One of the major issues was to establish if CAMPFIRE had incorporated indigenous knowledge in the planning and implementation of the programme. When asked how indigenous traditional knowledge is used in CAMPFIRE, a total of 110 (49 per cent) said that traditional and indigenous knowledge from Nyaminyami is not utilised at all in the CAMPFIRE programme. Another 43 (19 per cent) of the respondents do not know whether or not indigenous knowledge is used in CAMPFIRE. If indeed CAMPFIRE had utilised indigenous knowledge most of this group should have known, since it is their indigenous knowledge that would be reflected in the programme. This is highly significant, especially because the programme claims to incorporate local people's knowledge of conservation in order to make it relevant to them. The incorporation of indigenous knowledge was also meant to increase programme ownership by the local community, and indigenous knowledge was also supposed to provide some answers to problems of sustainable natural resource utilisation. In the CAMPFIRE literature many writers have suggested that one of the strengths of CAMPFIRE is that it is based on the use of indigenous knowledge and traditional practices. It is given as an approach that has adapted the traditional wildlife management system to present management needs. The results of this survey do not support this generalisation.

Only 6 per cent of respondents said that indigenous knowledge was used by CAMPFIRE in controlled hunting in three ways: the numbers

Table 7.1   How is traditional knowledge utilised in CAMPFIRE (1)?

Responses	Makande	Mola	Musamba	Nebire	Negande	Total
No. of interviewees	42	82	26	35	39	224
Utilised totems	5	0	1	0	0	6
Utilised hunting/ controlled	5	0	3	1	0	9
Not utilised at all	21	54	9	16	10	110
Don't know	7	13	0	9	14	43
N/A	2	0	7	3	2	14
Big difference	2	0	0	0	0	2
Consultations at meetings/elders	0	0	0	4	3	7
Advice and rules	0	0	0	0	4	4

Table 7.2   How is traditional knowledge utilised in CAMPFIRE (2)?

Responses	Makande	Mola	Musamba	Nebire	Negande	Total
Chiefs control animals killed	0	0	2	0	3	5
Not sure	0	5	0	0	3	8
Does not benefit community	0	0	1	0	0	1
Traditions are used in conservation	0	0	3	0	0	3
Avoiding fire forests	0	0	0	1	0	1
Communal ownership of animals	0	0	0	1	0	1
Utilised in learning from the past	0	10	0	0	0	10
Total	42	82	26	35	39	224

of animals killed, hunting seasons and in respect of totem animals. Whether clan names and clan animals do affect CAMPFIRE's operations was, however, heavily contested in the semi-structured interviews, where most respondents believed that CAMPFIRE allows the killing of elephants, buffaloes, sable and other totem animals. The elderly argued that CAMPFIRE does not encourage the killing of lions leopards or other so-called majestic cats, not because of the spirituality or totems attached to animals, but because they are considered majestic and in danger of extinction. CAMPFIRE's own values for protecting animals fit western values, but are not the same as those of local people and indeed are sometimes at variance with the local value system.

Another 10 per cent of the respondents believed that CAMPFIRE has learned from past traditional conservation practices and now utilises some of that knowledge in the programme. Indeed, there is evidence that CAMPFIRE has learned from some of those traditional practices as relates to calving seasons, movement of animals and even hunting seasons, but these influences seem to be minimal and have not significantly shaped the programme.

While a certain amount of consultation takes place with communities and chiefs, it does not necessarily reflect an adoption of indigenous knowledge by CAMPFIRE. The results of this survey suggest that CAMPFIRE does not recognise indigenous knowledge as contributing significantly to natural resource management. While CAMPFIRE acknowledges these as important in its documentation, in practice very little has been implemented. The study also found that:

1. In most cases in Nyaminyami today, indigenous knowledge exists only in theory: people can talk about it, but they do not have any practical experience, nor do they have a full understanding of what it is. This was particularly true of people who were 35 years or younger. There is a gap between knowledge and practice. Most of the Tonga people in rural areas have been alienated from these resources since colonial days and have never used these traditional practices because they have never had access to these resources. How then can they translate into action what they have never done?
2. Indigenous knowledge and traditional practices have been treated with nostalgic romanticism as if they exist independently of people's cosmovision. Indigenous knowledge cannot be translated into practice outside the phenomena of the cosmos. Many of the problems with indigenous knowledge in Nyaminyami emanate from this. CAMPFIRE has attempted to incorporate indigenous knowledge without understanding the people's cosmovision, nor accepting the Tonga spiritual and cultural values of biodiversity.
3. The dominant western development paradigm is based on conventional science, and that knowledge system has never understood

nor tried to understand issues of spirituality and the supernatural powers. CAMPFIRE is guided by this dominant development paradigm; how then can the programme incorporate that which it does not understand nor accept as reality. Conventional science does not give room to spirituality and spiritual growth, it is therefore impossible for CAMPFIRE to implement concepts from indigenous knowledge which it does not even believe in.

4. Many local people now treat their cultural traditions as inferior to the western cultural traditions, and hence are not in a position to take this any further than just talk about it. Their own knowledge has suffered serious erosion over time because it has been treated as inferior for so long. Most of the younger people now believe that there is no value to their own knowledge and have fully embraced western culture. They do not see how their indigenous knowledge can solve their problems of hunger, unemployment and illiteracy. They do not see how their indigenous knowledge can help them in a modern competitive world.

5. Western culture has always treated the African culture as inferior and primitive and has never allowed a genuine intercultural dialogue to take place. Since the western-trained ecologists at the Department of National Parks and Wildlife management who are responsible for CAMPFIRE have no capacity to handle issues of spirituality, how can CAMPFIRE incorporate indigenous knowledge?

6. Indigenous cosmologies and traditional practices have not always prevented over-exploitation of soils, overgrazing, deforestation and pollution, nor have they always promoted equity. Therefore, environmental degradation among the Tonga is not a new phenomenon. Many proponents of indigenous knowledge have treated the subject with nostalgia and as a novelty and have given an impression that going back to traditional practices and indigenous knowledge will solve all our problems. A very careful and serious assessment must be done of how much indigenous knowledge still exists, its quality and how much of it can still be practised, to determine which elements can be used. There are many traditional practices that were not environmentally friendly and therefore these practices cannot just be adopted wholesale.

7. Some indigenous knowledge and traditional practices have been adopted by CAMPFIRE. Local knowledge about the breeding of animals has been used, and hunting seasons have been adapted to the Tonga traditional seasons. The knowledge of local people is also

used when people come to hunt for sport; local people work as guides since they know the behaviour patterns of wild animals.

7. Conclusion

As indicated earlier, the issue is not whether or not indigenous knowledge contributes to conservation, but rather to understand the people's cosmovision and to use their yardstick to judge their contribution to sustainable resource utilisation. Opportunities have been lost in integrating conventional science with indigenous knowledge to benefit human beings and the environment. While indigenous knowledge has many advantages, not all practices can be adopted. Endogenous development will take place only when people's cosmovision directs their development. Endogenous development springs from indigenous knowledge systems and is inspired by the people's consmovision.

The foregoing evidence shows that most of the Tonga conservation ethic, indigenous knowledge and cultural and traditional practices are guided by spirituality. Those who come from outside this community are not able to deal with this spirituality. Unfortunately, there is no generally accepted theory that can explain or help us to understand the holistic approach or the laws of the spiritual world. Conventional science in its wisdom excludes spirituality and spiritual growth. The fact that conventional science does not understand spirituality does not mean that indigenous peoples have no conservation ethics nor should it demean spirituality.

Programmes such as CAMPFIRE face great difficulties in understanding the spiritual dimension of people's resource utilisation and conservation strategies. CAMPFIRE emphasises the interaction of the natural and the human world. It does not allow for the interaction between the spiritual world, human world and the material world. This is why CAMPFIRE is not an endogenous development and is unable fully to involve the people who live with and on these resources. CAMPFIRE is also unable to create an environment that allows for constructive dialogue between conventional scientific knowledge systems and the indigenous knowledge systems. Our ignorance should not be allowed to continue to degrade other knowledge systems. But above all the question can also be posed as to what is CAMPFIRE's conservation ethic if judged against Tonga spirituality?

The West needs to revisit its own spirituality when addressing the issues of sustainable utilisation of natural resources. The conventional scientific knowledge system cannot continue to call 'primitive' or 'unscientific' anything that this knowledge system does not understand. The fact that the West does not understand the indigenous people's conservation philosophy does not make that knowledge system inferior. The proponents of indigenous knowledge must stop being nostalgic but be more scientific in their search for what knowledge exists and what is still usable.

8
Problems of Intra- and Inter-group Equity in Community Forestry: Evidence from the Terai Region of Nepal

Rabindra Nath Chakraborty

1. Introduction

Community forestry – an institution of participatory natural resource management – is based on collective property rights to forests. As it restricts access to forests, it represents a particular solution to the problem of the commons (Gordon 1954; Hardin 1968; Ostrom 1990). Under community forestry, a local community is granted the right to extract forest products, apply silvicultural treatments and regulate access to a specified forest area. In this chapter I analyse problems of equity in community forestry with reference to the Terai region of Nepal. Community forestry in Nepal has experienced considerable growth in recent years: a forest law enacted in 1993 encouraged village residents to form forest user groups. The groups design and enforce a set of forest management rules in cooperation with the state forest administration. By December 1994, 2,756 community forestry user groups were managing 112,626 ha of forests (Hobley 1996: 89).

The rising interest in analysing the relationship between the internal structure of local communities and the outcomes of local natural resource management institutions has helped to refine the image of socially homogeneous communities who pursue a consistent, well-defined set of objectives.[1] The more sophisticated emerging viewpoint focuses on conflicts of interest within the group (Nadkarni *et al.* 1989; Ostrom 1990; Ostrom *et al.* 1994; Leach *et al.* 1997). As far as community forestry in Nepal is concerned, equity issues in the mountain regions of Nepal are being increasingly addressed (Chhetri and Nurse n.d.; Collett *et al.* 1996; Graner 1997). Here I present evidence

129

from the Terai lowlands, where community forestry is still in its initial stage.

The problem of equity in community forestry can be analysed on two levels. The first level – that of the user group – is appropriate for analysing equity within a forest user group (intra-group equity). On this level, two dimensions can be distinguished. The first is the distribution of forest benefits (such as forest products and the ecological functions of forests) across social actors. Here, however, I deal only with the issue of access to forest products, or an analysis of 'environmental entitlements' (Leach *et al.* 1997: 9). The second dimension is the distribution of the power to establish and change the rules of a user group (power in decision-making), which I treat as an indicator of the degree of participation.

The second level of analysis – that of inter-group relations – is useful for analysing problems of equity between a particular user group and other social entities in its environment (inter-group equity). Community forestry may, in practice, enable a village to monopolise control over a forest at the expense of others. The 'others' may be user groups in their formative stage or villages where no forest user groups exist. Again, two dimensions can be distinguished according to the distance at which the actors are located from the forest: first, groups who reside close to a forest (proximate users) compete with each other for access. Second, proximate users compete with users who reside far away from a forest (distant users).

The main result of this study is twofold. First, although all members of a community forestry user group benefit from community forestry in the long run, the scope for an active redistribution of decision-making power and forest benefits in favour of the disadvantaged sections of village society is limited because the stability of user groups is strongly based on the traditional structure of power in the villages. In this sense, a conflict exists between equity and user group stability. Therefore, the intra-group equity problem must be tackled by other, complementary strategies. Second, as far as inter-group equity is concerned, the scope for direct intervention is greater. However, it is limited by macro-level constraints.

The chapter is based on a field study in the Terai districts of Banke and Dhanusha, conducted during March and April 1997.[2] Banke District is located in the western part and Dhanusha is located in the eastern part of the Terai. Three to four forest user groups were visited in each district. Rapid rural appraisal methods were employed including group discussions, semi-structured interviews, participatory mappings and the drawing of forest lifelines (for a more detailed description of the methods employed, see Chakraborty *et al.* 1997: Annex 2).

2. Community forestry in Nepal

Forest policy in Nepal has undergone far-reaching changes during the last 50 years.[3] Until the 1950s, forests largely remained under traditional management systems. These systems restricted access by informal rules and thus, to a large extent, solved the problem of the commons. In 1957, forests were nationalised by the 'Private Forest Nationalisation Act'. Private ownership of forest land was abolished and all forests were declared state property (Talbott and Khadka 1994: 6). Access to the nationalised forests was restricted by the Forest Department, which was responsible for physically protecting forests against 'illegal' use by local populations. By the 1970s, it was gradually recognised that the Forest Department could not protect the forests effectively. As local people continued to depend on forest products for their livelihoods, they had no other option than to use forests illegally. As a result, the government made efforts to devolve the control over forests back to the local populations.

Community forestry in Nepal was started in 1978 when the Department of Forests handed over several government-managed forests to village *panchayat* committees.[4] Under these arrangements, the members of a forest user group comprised the population who lived within the boundaries of a particular *panchayat*. Problems arose because many forests were actually used by individuals who lived in different *panchayats*. In many cases, the *panchayat* committee members did not actually use the forest handed over and, hence, did not show much interest in its management. Finally, the legitimacy of the *panchayat* system of government was declining. As a result, property rights were insecure and participation was weak (Karki *et al.* 1994).

After the democratic movement transformed Nepal into a constitutional monarchy in 1990, the case for the transfer of control over forests to the local level was taken up again. A new Forest Act, passed in 1993, provides the legal basis for the implementation of current forest policy. Its provisions on community forestry can be considered as an effort to overcome the deficiencies of the former community forestry schemes. People who live in different administrative units (districts, Village Development Committees) are now allowed to form a single forest user group if they intend to use the same forest.

The transfer of property rights to a user group involves the following steps. When a forest user group is founded, it formally approaches the district forest administration for registration. Then a 'constitution' is drafted in cooperation between representatives of the group and the forest administration. In addition to the membership, organisational structure and objectives of the group, the constitution defines the distribution of forest products among the members. It contains a list of ordinary members and a list of members who perform administrative functions in the user group committee, its executive body. The constitution has to be enacted by the consent of all members and has to be endorsed by the District Forest Officer, the head of the district forest administration. The group can then apply for a forest to be handed over as community forest. To this end, a work-plan, negotiated between the user group and the forest administration, defines the geographical boundary of the forest, the silvicultural treatments to be applied, measures for forest protection, and the harvest of forest products.

Both the constitution and the work-plan have to conform to the rules defined in the Forest Act 1993, and the Forest Regulation 1995. For example, forest land can never be converted to non-forestry uses, but there is generally considerable scope for designing the work-plan according to local needs. The negotiating power of the village residents in this process is limited, however, as the forest user groups are unable to exert any significant pressure on the District Forest Officers. After the work-plan has been endorsed by the District Forest Officer, the property rights defined in the work-plan are transferred to the group. If the rules of the work-plan are violated, the district forest administration can decide to withdraw the rights.

An underlying intention of contemporary community forestry legislation in Nepal is to restore traditional common property forest management regimes in a modernised form. Individuals who historically used a forest are encouraged to establish a user group in order to be granted collective property rights to that forest. In the Terai region, however, it is hardly advisable to restore the status quo because the overwhelming part of the present population migrated to the region after 1960. Therefore, different criteria for user group formation and the transfer of property rights to forest user groups have to be applied to the Terai. The forest administration has responded to this situation in two ways: one is to delay the transfer of forests until these conflicts are resolved; the other is to hand over the forests to the first claimants.

3. Community forestry in the study area: an overview

3.1   Forest cover and location of the groups visited

Differences and commonalities exist between the two districts.[5] In Banke, an area of 148,111 ha (66 per cent of the district) is covered by forests, which are easily accessible and commercially valuable. In Dhanusha, by contrast, forest cover is only 30,846 ha (26 per cent of the district), with almost all the remaining natural forests concentrated in the north of the District. Access to the Dhanusha forests is difficult, as they are located in the Shivalik Hills. Estimates on deforestation rates are available only for the plains in the two districts. While the annual rate of deforestation in Banke was 1.18 per cent during 1978–90, forest cover in the plains of Dhanusha increased by 2.2 per cent annually (His Majesty's Government and FINNIDA 1994: 6). This increase – from a very low level of 2,000 ha in 1978 – has been attributed to tree planting on private land.

The location of the villages visited is as follows. In Banke, the villages and community forests of three forest user groups (Shrideshwor Mahadev, Binauna, Mahila Upakar) are adjacent to extensive natural forests. The fourth group (Gijara), in contrast, manages a 100 ha community forest that is distant from the natural forests. In Dhanusha, two user groups and one other village visited are located close to the natural forest in the north of the District (Madhubasa, Kemalipakha, Maal Tole) while one user group (Haththipur) manages a small (5.5 ha) canalside plantation forest in the centre of the District.

3.2 User group formation

The rural population of the Terai region depends on forest products to a considerable extent. Firewood accounts for 67 per cent of their energy consumption.[6] Furthermore, timber is used for housing and agricultural implements. Grass and tree leaves from the forests are important sources of fodder. Therefore, many agriculturalists send their animals to the forests for grazing. Finally, rural households collect a large spectrum of other non-wood forest products such as fruits, flowers, roots, leaves and mushrooms (Chakraborty *et al.* 1997). No pastoralists live in the study districts.

Community forestry is still in its initial phase in the two districts, yet all user groups visited (except one) have been highly stable: the members have complied with the rules for an extended period of time, mainly because community forestry builds on the existing village power structures and 'systems of authority' (Bromley 1992: 9). Moreover, they reinforce existing patterns of power and authority, as powerful community members increase their influence through their control over the forest as an important resource, as can be seen from a brief overview on the state of user group formation in the two study districts.

3.2.1   Banke District

As Table 8.1 shows, only four community forests have been handed
over in Banke District so far, three of them during the fiscal year
1996–7. The total forest area handed over to local user groups – about
200 ha – represents only a tiny fraction of the District's total forest area
of 148,800 ha. Up to now, the forest area handed over to community
forestry user groups consisted either of forest plantations (Gijara) or
degraded natural forest (Srijana, Mahila Upakar and Rimna). No well-
stocked forest was handed over.

The forest user groups comprise 548 households in total. Forest area
per member household is similar in all groups except Rimna User
Group. Nine more forest user groups have not yet been transferred any
property rights. Some of them have just been established while others
have already prepared their constitutions. The survey of the area of the
proposed community forest has been completed with two forest user
groups. Community forestry user groups are not only being formed in
easily accessible areas (e.g. close to the highway or to forest range
posts) but also in remote areas (e.g. east of the Rapti River).

People are aware of the importance of forests and the benefits of
community forestry. As a consequence, several villages (such as
Binauna) have developed informal protection systems in several locations and protected the forest for several years even without being offi-
cially registered by the District Forest Officer. Another example is the
Gijara User Group. After a plantation had been established by the forest
administration close to the village of Gijara, awareness grew that the

Table 8.1   Community forests transferred, Banke District

Date of transfer	Name of user group	Location (name of village)	Forest area (ha)	Number of households	Forest area per member household
1995–6	Gijara User Group	Udarapur	134	256	0.52
27.2.1997	Srijana User Group	Kohalpur	10	26	0.38
20.3.1997	Mahila Upakar User Group	Kohalpur	26	63	0.41
1997 (planned)	Rimna User Group	Mahadevpuri	30	203	0.15
Total			200	548	0.36

Source: Data provided by the District Forest Office, Nepalgunj, Nepal.

forest might be handed over to the community some day. As a result, the plantation was protected for four years before the forest was actually transferred to the user group.

3.2.2 Dhanusha District

In Dhanusha District, 17 groups have applied for the transfer of community forests, but the only group who has been handed over the forest so far is the Kamal Nahar Forest User Group in Haththipur. The other groups – all registered in 1996 – are still in the process of drafting (and finalising) the constitution and the work plan (see Table 8.2). Because natural forests are all found in the north of the District, most community forests are situated there. Only three forests are located further south (Hardinath User Group, Kamaladhar User Group and Kamal Nahar User Group). The area of the community forests (as defined by the Forest Department) varies from 1 ha (Tallo Chhaghariya Mahila User Group) to 200 ha (Saphee Inar User Group). The area of all registered (prospective) community forests is 1,083 ha, which amounts to 3.5 per cent of the total forest area in the District in 1997 – but only 5.5 ha (the Kamal Nahar community forest) have actually been handed over so far, and the areas registered for the other community forests may not actually be transferred, since the District Forest Office can decide to meet only part of the claim. The variation across groups in forest area per member household is much greater than in Banke District. Forest area per member household ranges from 0.02 ha (Kamal Nahar User Group) to 1.42 ha (Lalpur User Group). However, the *average* community forest area per member household is similar in the two districts: 0.36 ha for Banke and 0.41 ha for Dhanusha, and the District Forest Officers may be following unofficial or informal norms in defining the size of a community forest.

3.3 Conflicts with the forest administration

The field visits revealed that further demand for community forestry, beyond the abovementioned groups, exists in the two districts. Forest user groups say that the forest administration delays the process of registration and handover. The forest administration, in contrast, insists that it must carefully examine the definition of group membership because the handover of a forest cannot be based on traditional rights as in the mountains. Some parts of the forest administration may, however, have an interest in limiting (or at least delaying) the transfer of forests to user groups. Forest rangers, for example, receive bribes to allow village residents to collect firewood in government-managed

Table 8.2 Registered forest user groups in Dhanusha District

Date of registration	Name of user group	Location (name of village and settlement number[s])	Forest area (ha)	Number of households	Population	Forest area per member household (ha)
28.6.1996	Madhubasa User Group	Pushpalpur – 9	77.50	61	344	1.27
2.7.1996	Kharshange Danda User Group	Bengadabar – 3, 4	52.64	201	801	0.26
9.12.1996	West Kharshangedanda User Group	Bengadabar	85.04	176	839	0.48
30.6.1996	Budha Shanty User Group	Hariharpur – 7	25.00	148	870	0.17
5.7.1996	Shidh Shanty User Group	Hariharpur – 9	22.00	93	491	0.24
7.7.1996	Kemalipakha User Group	Dhalkebar – 1	150.00	141	741	1.06
14.11.1996	Pokhari Damar User Group	Dhalkebar – 2–5	182.00	390	2,300	0.47
7.7.1996	Saphee Inar User Group	Bharatpur – 2	200.00	198	1,187	1.01
7.7.1996	Lalpur User Group	Bharatpur – 2	75.00	53	296	1.42
8.7.1996	Aaurahi Baba User Group	Naktajhej – 9	150.00	186	1,127	0.81
8.7.1996	Tallo Chhaghariya Mahila User Group	Bhartapur – 2	1.00	36	229	0.03
8.7.1996	Kamala User Group	Lovtoly – 4–8	17	236	1,261	0.07
9.7.1996	Quarter User Group	Yagyabhumi – 7, 9	9.43	58	303	0.16
6.9.1996	Hardinath User Group	Gopalpur – 8–9	8.00	159	1,105	0.05
9.7.1996	Kamaladhar User Group	Paterba – 1–4	23	282	1,744	0.08
9.12.1996	Kamal Nahar User Group	Haththipur – 6–9	5.50	256	1,490	0.02
Total			1,083.11	2,674	15,028	0.41

Source: Data provided by the District Forest Office, Janakpur, Nepal.

forests. If a forest is handed over to a community forestry user group, this will relieve pressure on the government forests and, hence, reduce forest rangers' incomes from bribes in the long run.

As a result, conflicts with the forest administration during the handover negotiations are frequent. In Maal Tole Village (Dhanusha District), the village residents stopped informal protection of a nearby forest because they felt that the district forest administration did not really intend to hand over the forest to them. In the Kamal Nahar User Group, the only group who has been handed over a forest in Dhanusha so far, the handover process was successful because a former village resident (now working as an officer in the Department of Soil Conservation) initiated the formation of the user group and assisted the village in pursuing the transfer vis-à-vis the forest administration.

4. Intra-group equity

What is the impact of community forestry on intra-group equity? Intra-group equity is conceptualised as the distribution across classes of actors – differing with regard to income, ethnicity and gender – of benefits from the community forest and of power in decision-making.

4.1 Income equity

4.1.1 Distribution of forest benefits

In considering problems of equity between income classes, the concept of 'income' must encompass both cash and non-cash incomes, the latter valued at market prices. The central equity problem with community forestry stems from the fact that most forests are strictly protected in the short run, which hurts the poor more than the non-poor. In the long run, however, the poor gain in absolute terms.

In the long run, every member of a forest user group benefits from community forestry, since otherwise the accessible forest area would further decrease and finally vanish. Degraded forests that are temporarily closed and effectively protected will yield more forest products in the future. In this sense, community forestry leads to a Pareto improvement for the members of a user group.

Will the long-run gains of forest protection be distributed equally across income classes? In many user groups, appropriation rules do not discriminate against the poor. No member is formally excluded from using the community forest, if forest use is permitted at all. Conversely, if a forest is closed, it is closed to everybody. Two factors, however, tend to shift the distribution of income (measured by the relative income share of the poorest 50 per cent of the user group members) against the poor: collection fees, and the opportunity for the non-poor to appropriate disproportionately high benefits from the forest. Collection fees are charged in certain (but not in all) groups. In the Gijara forest user group, for example, all members are charged collection fees for extracting timber (Rs 15 per day), firewood (Rs 25 per headload per day) and thatching grass (Rs 5 per headload per day). Members who cannot pay the fees are not allowed to harvest these products. In Madhubasa, firewood collection is charged at Rs 5 per bundle per day. These fees are substantial given the (male) agricultural wage level of Rs 40–60 per day. These fees may lead to the *de facto* exclusion of the low-income classes, but our respondents stated that all user group members were prepared to pay the fees.

Collection fees have a bias against the low-income classes even if the poor are willing (and able) to pay. If forest products are collected for self-consumption, the collection fee represents consumption expenditure. In this case, the rich are able to collect more forest products than can the poor because their income and, hence, their consumption budget is higher. If the forest products are collected for market-oriented production, it may be possible to pass part of the fee to the customers. In this case, the fee represents an investment in working capital, which can be financed by credit. As the rich have easier access to credit than the poor, the bias exists even in this case.

In other groups, the rules allow the non-poor to appropriate disproportionate benefits. For example, if each family is allowed to graze their animals in the forest, a rich family owning many cattle can appropriate more fodder grass or leaves than a family owning only few animals. More generally, if the appropriation of common property resource units requires the application of private inputs, actors who are better endowed with private inputs can harvest more resource units than those who are less endowed.

One factor tends to shift the income distribution in favour of the poor. A poor household generates a higher *fraction* of its income from a community forest than a rich household does, because it is less endowed with private factors of production. If the level of income depends incrementally on the productive use of private assets and the appropriation of resource units from a common property resource, a given and identical absolute increase in the rate of appropriation of resource units for all actors induces a disproportionate increase of income in the hands of the actors who are less endowed with private assets. As a result, the income distribution shifts in favour of the poor.

An example is energy consumption. A household that owns less land is more reliant on fuels from common property sources than a household that owns more (Soussan *et al.* 1991: 1305). Landless labourers represent an extreme case; many of them depend totally on common property resources for their self-consumption of firewood, fodder grass and other non-wood forest products.

The effect of community forestry on the distribution of income in a user group thus depends on the relative strength of the three forces just described. A slight increase in the income share of the poor can be expected if collection fees are absent and each family can appropriate the same amount of benefits in absolute terms.

In the short run, the income distribution shifts against the poor as a result of community forestry. As most community forests are initially closed for protection, the poor – especially the landless and members of particularly forest-dependent professions – must either purchase wood and fodder or illegally extract it from government-managed forests. The rich, in contrast, can use their private assets to produce forest products. Several landowning respondents reported that they increased tree planting on their own land in response to the reduced availability of wood from the community forest.

Both in Banke and in northern Dhanusha, the overwhelming majority of the user group members rely on government-managed forests for fuelwood extraction. Only in one case (Gijara Forest User Group) was the amount extracted from the community forest sufficient to cover the fuelwood demand of the group, for two reasons. First, most community forests that have been handed over so far are heavily degraded and require protection for a substantial period of time. Second, wood extraction rules are very restrictive, particularly, never allowing the cutting of living trees. On one hand, this is a simple rule that is understandable to everybody and can be monitored at a low cost. On the other hand, both foresters and local communities perceive a problem of overprotection (or underutilisation) in some locations. Wood output could probably be increased by silvicultural treatments involving the cutting of green trees, for example, thinning and rotational fellings.

Members of strongly forest-dependent professions are a group within the landless (or near-landless) households whose welfare is very strongly affected by the closure of a community forest. Examples are blacksmiths or bamboo-basket weavers. A basket weaver family in Maal Tole, where no community forestry user group has yet been founded, strongly expressed their opposition to community forestry. The man spends one full day a week collecting bamboo, which is their raw material. His wife collects firewood for cooking twice a week, leaving around 9 a.m. and returning around 3 p.m. Both expected that their collection times would increase if the nearby forest was declared a community forest and protected. In contrast, the better-off sections of the village strongly favoured the establishment of a community forest.

4.1.2   Distribution of power in decision-making

Participation in decision-making on community forestry management rules is limited. Village leaders, who usually belong to the wealthy strata of the community, support community forestry and determine the rules for the user group together with the rangers. As a result, the actual process of the formation of the user group committees remains unclear to us even after many interviews. User group committee members are not, apparently, elected in a true sense. Instead, they are presented (or present themselves) to the general users meeting and are assigned their positions by an act of general consent (acclamation). This is reflected in some of the user group constitutions. The Mahila Upakar User Group, for example, specifies the following procedure for the appointment of committee members:

> To implement the work plan prepared by the group efficiently[,] a committee is formed. Members of the committee will be nominated from the general assembly of all the users with equal representation from the [*sic*] village and caste group. (Mahila Upakar Community Forestry User Group 1996)

In most villages, the less powerful users did not object to these rules when we talked to them. This evidence has to be interpreted in the light of the general result reported above that user groups build upon the existing power structures in the villages. The poor seem to accept their (weak) position rather than try to use community forestry as an instrument for redistribution.

4.2   Equity between ethnic groups

The inhabitants of the Terai region can be divided into three broad ethnic groups: the *Tharu*, mountain migrants and the Terai castes. As malaria was endemic in the region before 1960, the *Tharu* – who were immune to malaria – were the only ethnic group to settle there. Their economy was based on swidden cultivation and silvipastoralism; land was generally managed under common property. After 1960, public malaria eradication programmes and resettlement schemes stimulated a rising flow of migration into the Terai. The migrants were able to impose on the Tharu a structure of land property rights based on private property, which resulted in a process of land alienation from the Tharu to the migrants. At the same time, the migrants applied a superior (intensive) agricultural production technology to the land, which reinforced the shift in the income distribution in their favour. As a result, the Tharu were marginalised both economically and socially (Müller-Böker 1995: 96, 162). The migrants came from two regions: the mountains and the Indian part of the Gangetic plain. Mountain residents frequently migrated to the Terai in groups – in some cases, entire villages – from the same region (e.g. Madhubasa village). As far as migrants from India (Terai castes) are concerned, their exact place of origin is not always clear.

With regard to the ethnic composition of forest user groups, two patterns can be identified. On the one hand are homogeneous user groups, consisting either of mountain ethnic groups (Mahila Upakar, Shrideshwor Mahadev, Madhubasa, Kemalipakha) or Terai castes (Haththipur); on the other, mixed groups, consisting either of mountain ethnic groups and Tharu (e.g. Binauna) or of mountain ethnic groups and Terai castes (Gijara and Maal Tole).

In the distribution of forest benefits, harvesting rules do not discriminate against particular ethnic groups, but equity effects arise to the extent that ethnicity is linked to income. As the Tharu are generally poorer than the mountain migrants, they are more strongly affected by the distributive impact of community forestry (see section 4.1.1) than the mountain migrants.

Discrimination appears to exist with regard to decision-making. The mountain migrants, who occupy the strongest position in the hierarchy of ethnic groups, have the greatest influence in defining the constitution and the work plan in a mixed user group. Apart from the factors mentioned already, the centre of political power in Nepal has been situated in the mountains for more than two centuries, and the dominance of mountain migrants vis-à-vis the Tharu and the Terai castes reflects this more general feature of Nepalese society.

4.3 Gender equity

In the long run, that is, after the reopening of a community forest, *women* will benefit strongly and directly through reduced collection times for forest products, but their *total* workload may not decrease if their working hours increase in other areas. *Men* in turn benefit both indirectly and directly from community forestry. They benefit indirectly if the time saved by women in connection with forest use is reallocated to agricultural or other work. They benefit directly if they collect fuel-wood, timber or other forest products themselves. In the short run, however, the closure of community forests for protection has a gender bias, as the tasks of collecting wood and grazing animals are tradition-ally performed by women. For women in poor households, this adds to the bias against the poor described above.

Women participate less than men in the decision-making processes, though every committee visited has one or more female members. However, the ethnic composition of a user group is a significant factor. Women's position is comparatively stronger in groups consisting exclu-sively of migrants from the mountains, and is very weak in groups con-sisting solely of Terai castes. One possible reason is that the institution of *parda* (seclusion of women) is much stronger with the Terai castes than in the mountain communities. In the exclusively Terai caste Haththipur Forest User Group (Dhanusha District), the three women in the user group committee were unable to explain the management rules to the research team although they regularly attend the meetings. At the same time, they strongly favoured an increase in the number of female committee members, which indicates that they cannot actively participate in the meetings because their position is weak.

5   Inter-group equity

Problems of inter-group equity can occur among proximate users and between proximate and distant users of a forest. Proximate users reside less than half a day return trip from the forest where they intend to harvest forest products. When several villages compete for the use of the same forest, forming a community forestry user group may enable a relatively prosperous village to monopolise control over the forest at the expense of other, relatively poor villages. Similarly, there is an equity conflict between proximate and distant users because commu-nity forestry enables proximate users to assume control over forests at the expense of distant users.

5.1   Proximate users

5.1.1   *Income equity: prosperous* vs. *poor villages*

In many cases, prosperous villages are more active and articulate in pushing forward their demands for the handover. They are better informed about community forestry and are fast enough to be the first to gain control over the resource. Among the groups visited, this applies specifically to villages with a majority of mountain migrants.

There appears to be a link between economic prosperity in a village and success in community forestry.[7] Prosperous villages produce an economic surplus sufficient to enable the residents to allocate time to collect information about community forestry, lobby the forest administration for handover, and spend time on the intra-group processes required for effecting a handover.

Lobbying takes considerable time. As bribes paid for entering government-managed forests are an important source of income for the forest rangers, the rangers are reluctant to support the conversion of government-managed forests to community forests, and they delay the handover process as much as possible. As a common response, village leaders try to bypass the rangers by directly contacting the District Forest Officer. This creates substantial transaction costs, as they must travel to the district capital and wait for many hours in the District Forest Office. Poor villages can rarely afford this time. Negotiations with the forest administration on the boundaries of the forest and the rules for forest use and protection also take time. Several groups visited complained that meetings with rangers in the villages fail because the rangers do not appear at the agreed time.

The experience of the Gijara forest user group provides an example of inter-group equity conflicts between proximate users. Mountain migrants from the Pokhara region founded the village of Gijara in 1973. A large area of forest was probably cleared at that time for agricultural use. In 1989, the government initiated a tree-planting programme, and 100 ha of degraded land were planted with trees (*sissoo, babool, sal* and *pipal*). There were three groups of tree planters: the village residents, who are now members of the user group; residents from adjoining villages, not members of the user group; and a group of young men from outside the area who had been gathered by the government as volunteers.

In the beginning, the residents from all villages were suspicious about the project, as they feared that the forest administration was interested in making them contribute labour without making the benefits available to them. When the trees grew up, however, the residents of Gijara village realised that they could appropriate the benefits and started to protect the forest. The 'protection' activities involved preventing the residents from other villages from extracting forest products, leading to a conflict with the other villages. One respondent stated that the residents of Gijara 'caught' residents from other villages 'stealing' thatching grass (*khar*) from the forest. The grass was confiscated and sold. As the conflict increased, Gijara residents turned to the forest administration, which recommended them to found a forest user group. The group was constituted and the forest handed over in 1996.

Of course, the conflict with the adjoining villages persisted because the community forest is the only forest within a distance of several kilometres. As a result, a compromise emerged. The residents from the adjoining villages can now collect forest products under the same conditions as the Gijara residents, that is, against the payment of a collection fee. Furthermore, the forest user group contributed money to the construction of a building for the local school, attended by children from all villages, easing the tension considerably.

The Gijara case shows that community forestry can enable prosperous communities to monopolise control over a forest at the expense of other, poor communities. However, the leadership role a village performs in community forestry is not necessarily detrimental to other villages. In some cases, the transfer of exclusive property rights to a village under community forestry motivates the residents of other villages to form user groups themselves and apply for being handed over another forest, as with the Kemalipakha forest user group in the north of Dhanusha District.

In Kemalipakha, increasing scarcity of wood and other forest products led to a decision by a general village meeting in 1982 to protect a forest. In 1990, after eight years of informal protection, a forest user group was formed. Forest protection meant that access to the forest was restricted not only for the Kemalipakha residents but also for adjoining villages, whose residents objected. In this phase, the district forest administration strongly supported the claim of the Kemalipakha residents on their community forest and tried to convince the other villages to establish their own community forests. As a result, several villages established their own community forests in 1994. In several villages more than one user group has been constituted (Bengadabar, Hariharpur, Dhalkebar and Bharatpur) (see Table 8.2), indicating that the formation of a user group in one settlement motivates the residents of other settlements of the same village to form user groups themselves.

5.1.2   Mountain migrants vs. Tharu

In some cases, particular ethnic groups are excluded *de facto* from forest use. For example, the Mahila Upakar community forestry user group near Kohalpur, Banke District, was founded by women members of a non-governmental organisation for the advancement of women (Mahila Upakar) that supports small-scale economic activities (weaving, bread-baking, poultry). The forest area (26 ha of heavily degraded land) was handed over in 1997 and will yield forest products only after several years of absolute protection. The members established a nursery and planted saplings on the community forest area. The members live scattered over an area that comprises several villages. Membership in the non-governmental organisation is compulsory for the user group members. As the members feel that substantial pre-investments have been undertaken in their non-governmental organisation, they charge an admission fee of Rs 2,500 to new members. The fee clearly discourages the poor from applying for membership, as the daily wages for agricultural labour are Rs 20–30 for women and Rs 50–60 for men. Several members of the Tharu communities who reside in this area complained about being denied the opportunity to enter the group. Even the current membership fee of Rs 10 per month, which is used to pay the salary of the forest guard, is too much for them. Several Tharu women who were members of the NGO in former times left the organisation recently. The remaining members of the NGO and the forest user group are mountain migrants.

5.2   Distant users

In Dhanusha, the distant users live in the southern and central parts of the District. In Banke, they live in the southwest. As their villages are scattered over a wide area, they cannot organise themselves and form a user group. The way distant users adjust to fuelwood and fodder scarcity strongly depends on their land tenure status. At one extreme of the spectrum of tenure arrangements, *landowners* grow trees on their own land and substitute animal dung and agricultural residues for fuelwood. At the other extreme, the *landless* have to rely on common lands. With increasing distance to natural forests, common lands become smaller and carry less trees or grass. In two villages visited in the south of Dhanusha District, the size and productivity of the common lands were clearly insufficient to cater to the needs of the landless. A common response is that the landed classes tolerate the landless' stealing stalks and twigs from private land. However, this 'tolerance' is not really altruistic, as the landed classes occasionally demand labour services from the landless in exchange.

6.   Implications for policy

6.1   Intra-group equity

The evidence encountered in the user groups may be summarised as follows. Although in the long run, the distribution of benefits from the forests may shift the income distribution in favour of the poor, in the short run the poor and the women have to suffer disproportionate losses. Moreover, the scope for participation in decision-making on user group rules is weak for the landless, women and disadvantaged ethnic groups such as the Tharu.

At the same time, a high degree of stability could be observed in all but one of the user groups visited. This gives rise to the following hypothesis. The general distribution of wealth and power in a village is largely determined by the distribution of the ownership of productive assets, especially land. The relative congruence between the distribution of decision-making power and benefits from forestry *within a user group* and the general distribution of wealth and power *within the respective village(s)* is a major reason for the observed stability of community forestry user groups.

This has implications for the scope of direct interventions that aim at changing the rules of decision-making and the distribution of forest benefits in favour of the disadvantaged sections of village society. Small changes in this direction may be successful, but will have only a small impact on the distribution of wealth and power within the user group. If big changes to the rules are effected by legal provisions, pressure from the forest administration, or moral suasion by donor agencies, an incongruence will develop between the distribution of power in the user group and the distribution of power in the village. This incongruence will threaten user group stability. In this sense, a trade-off exists between the objectives of equity and user group stability.

As a consequence, the scope for an *active redistribution* of benefits appears to be limited. For example, the short-run equity impact of the closure of the community forests could be mitigated if the landless poor and members of especially forest-dependent professions were allowed to collect small quantities of forest products, even during the period of strict protection. Such a rule is unlikely be adopted, however, given the power structure in the user group committees. If it is forced upon a user group (for example, by law), the disadvantaged sections of the user group may well not actually benefit from such provisions. Instead, the rich may present themselves as the 'neediest'. Alternatively, the non-poor may decide to extract forest products, too. This would amount to a breakdown of forest protection or, in other words, to the destabilisation of the user group. Similarly, the poor are unlikely to be allowed to appropriate higher absolute benefits per family than the non-poor in the long run. In this sense, the impact of community forestry on *poverty alleviation* is limited.

We should not, however, assess the impact of community forestry on poverty alleviation exclusively by comparing the present reality with a desired state of community forestry. From a natural resource management perspective, it is equally important to compare the effects of community forestry with the effects of alternative forest management institutions, for example forest management by the state or by private actors, and community forestry performs well in this respect (Chakraborty *et al.* 1997).

Three classes of measures could be taken to improve the position of the disadvantaged sections. First, to increase the representation of the landless, women, and marginalised ethnic groups in the committees. This would involve small changes in the rules, not threatening stability but enabling the disadvantaged to pursue their interests more vigorously. At the same time, it would raise the confidence of these groups in their own capabilities. Second, to raise the potential yield of forest products. An obvious strategy is to hand over not only degraded forests but also well-stocked forests. This would allow a positive level of extraction in the well-stocked part of the community forest while the degraded parts are protected. However, this may not be feasible in all locations (see below). A third policy option is to improve the situation of the disadvantaged sections through complementary measures. These could include rural credit programmes or, more generally, programmes for income-generating activities targeted at the landless, women, or marginalised ethnic groups. More indirect measures are policies that promote agricultural growth. Agricultural growth raises the demand for labour in agriculture and in the non-agricultural sectors (Rao and Caballero 1990). The strongest factor in promoting structural change in favour of the weaker sections of village society is the growing scarcity of their labour, reflected in rising rural wages and rising incomes of the self-employed.

6.2   Inter-group equity

The inter-group equity problems mentioned above are more amenable to direct intervention. However, these options are limited by macro-level constraints. The state forest administration has an important role in resolving distributive conflicts between the proximate users of a forest. The District Forest Officer could ensure that all villages close to a forest are informed when the forest is considered for handover. Furthermore, (s)he could propose including several competing villages in one user group, enabling the administration to prevent the monopolisation of a forest by a single village. NGOs could equally perform the task of informing villages. Several patterns of co-operation between NGOs and the forest administration are possible in this context.

Conflicts between the interests of proximate and distant users are more difficult to resolve. One option for the state is to reserve part of the forest area in a particular region for resource extraction by distant users. However, there is a targeting problem because the general opening of the forest will also attract proximate users. A second solution is to develop markets for forest products further, enabling distant users to purchase the forest products they need from community forestry user groups. This 'solution' cannot be considered as satisfactory, however, because the distant users who are most dependent on forests – that is, the landless – cannot afford to buy them.

A third option is to make proximate users (that is, community forestry user groups) share their property rights with distant users, for example, on a seasonal basis. This could, however, give rise to legitimacy problems if the distant users appropriated forest benefits while all the costs of forest protection work fell on the proximate users. A solution could be to compensate the proximate users for their cost, but integrating distant users into community forestry user groups might make the institutional arrangements more complicated and, hence, more fragile.

A general limit to the capacity to resolve inter-group equity conflicts is that there may not be enough forest resources in a particular location to satisfy the resource needs of both proximate and distant users. In this case, there are several options. First, the output of wood could be raised by appropriate silvicultural treatments (His Majesty's Government and FINNIDA 1993). Second, the dependence on forests as a source of energy could be reduced by a shift towards alternative energy conversion technologies, for example, more efficient stoves, biogas, solar cookers or commercial fuels. However, the adoption of most of these options presupposes an increase in income, which, in the long run, will have to be achieved by agricultural and industrial growth.

Notes

1. See, for example, Shiva (1991: 272–3).
2. The field study, which was part of the 1996–7 training course at the German Development Institute, Berlin, was carried out by Ines Freier, Friederike Kegel Martina Mäscher and the author. I thank the other members of the study team for their cooperation and numerous constructive discussions.
3. See Talbott and Khadka (1994) for a brief but fairly comprehensive account.

4. The *panchayat* (council) system was Nepal's form of government before 1990. At the lowest administrative level, the village *panchayat* was the body who governed a village or a group of settlements. At the same time, the concept of village *panchayat* referred to a territorial entity, that is, the territory governed by this body. The *panchayat* system was replaced in 1991 with the introduction of a multi-party parliamentary system. The Village Development Committee (VDC) is now responsible for village affairs.
5. Information supplied by the district forest administration in each case.
6. Calculations based on Water and Energy Commission Secretariat (1996: Annexes 1 and 4.1-c).
7. If 'poor' and 'prosperous' villages are defined in terms of per capita wealth, a poor village could be more successful in community forestry than a prosperous village because it has a small group of wealthy and determined leaders who pursue matters vigorously. However, this did not appear to be typical in the villages visited.

9
Benefits to Villagers in Maharashtra, India, from Conjunctive Use of Water Resources

Frank Simpson and Girish Sohani

1. Introduction

This chapter presents results of and lessons learned from a multidisci-plinary research project, 'Conjunctive Use of Water Resources in Deccan Trap, India'. The project involved Bharatiya Agro Industries Foundation (BAIF), Pune, Maharashtra State, India, and University of Windsor (UW) Earth Sciences Department, Windsor, Ontario, Canada, working in partnership with people from Akole Taluka, Ahmednagar District, Maharashtra State, India. Participatory management and eval-uation were key elements of the project (Sohani *et al.* 1998).[1] Planning of the project was finalised when IDRC brought the intending partners together in the summer and autumn of 1991. The project term was ini-tially three years, beginning in April 1992, but later was extended until March 1996. Unspent funds were reallocated for a field assessment of project sustainability in April to May 1997.

The project purpose was to design a management strategy to provide a year-round water supply in selected villages of Akole Taluka (Figure 9.1). The initial intention was to concentrate on problems of domestic sup-ply. However, efficiency of water use required a more holistic approach, including agricultural uses of water. The evolving strategy also came to include construction of demonstration sites for several appropriate water conservation technologies.

During the driest part of 1997, less than a month before the onset of the monsoon, villagers had abundant water for use: in deepened, dug wells; in reservoirs, occupying the valleys of ephemeral streams; and in tanks, trapping the spring waters. Agricultural production also showed

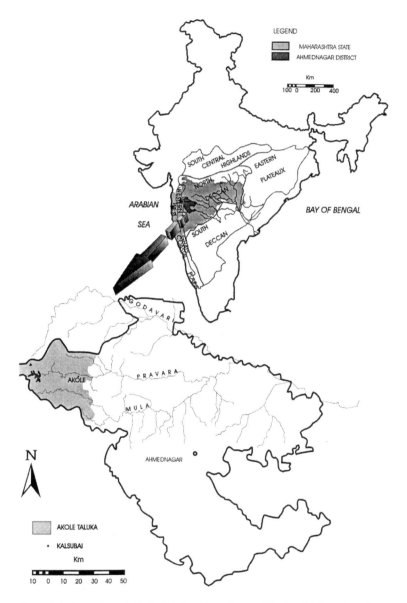

Figure 9.1 Location of Akole Taluka, Ahmednagar District, Maharashtra State, India

clear benefits from the added soil moisture retained in the fields, and from a related reduction in soil erosion.

The project was a success: an effective water resource management strategy was in place. Indeed, the demonstration sites were effective to the point of satisfying the water needs of the villagers. In this chapter we outline stages of the project life cycle, with special reference to the main contributions made by local people. We also describe the principal benefits they received from the project, as well as related impacts on village life.

The objectives of sustainability, effective use and replicability, as well as capacity-building in support of them, are important elements of projects that focus on the provision of water supply and sanitation through participatory management (Narayan 1993). A unique characteristic of this project was the close interaction between the villagers and BAIF's field officers, which permitted consideration of these elements on a daily basis and facilitated community participation in all stages of the project.

2. Problems

Akole Taluka is in the northwestern part of Maharashtra, on the eastern flank of the Western Ghats mountain range, in the high plateau that gives topographic definition to the Deccan Trap province of flood basalts over more than 500,000 sq km of western and central India. Kalsubai (1,646 m), the highest mountain in the Western Ghats, is near the western margin of the *taluka*. Relatively steep slopes and rugged topography in the west are replaced by more gently undulating and flat land further to the east and south. Project activities were concentrated around the villages of Ambevangan, Manhere and Titvi, in the west-central part of the *taluka* (Figure 9.2). Surface drainage in this area is generally southwards into the east-flowing Pravara River, a major tributary of the Godavari River, by way of numerous, mainly ephemeral streams.

Soils in the area are most extensively developed in the relatively wide, lower reaches of the larger valleys, where thicknesses of several metres occur. They rest on weathered bedrock, generally up to 2 m or so thick, and basalt lavas that are the bedrock. Rock exposures in the project area are small and scattered. By analogy with observations in adjacent areas, the lava flows are considered to be generally flat-lying and laterally continuous over thousands of metres. The lavas are impermeable' and movement of ground water in them is largely

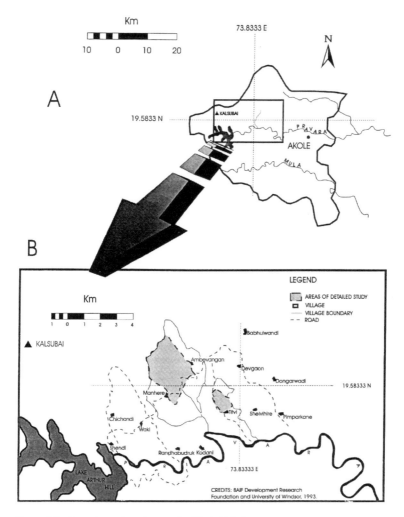

Figure 9.2 Location of study area in Akole Taluka

restricted to discontinuities between flows and to fractures that sporadically interrupt the lateral continuity of flows. Somewhat improved aquifer conditions exist in the soils and weathered bedrock, as well as in the sediments occupying the channels of larger streams.

Akole Taluka receives rainfall from the southwest monsoon between June and September. July is the wettest month. Rainfall across the area ranges from 2,000 mm in the west to 600 mm in the east. In the past, the monsoon period was characterised by widespread transfer of water and eroded soil to lower elevations by surface runoff. During the post-monsoon period of October through to January, water supplies in close proximity to the villages would dwindle. The flow of streams gradually decreased, many springs showed reduced discharge and water levels in wells dropped. The pre-monsoon period of February through May saw a progressive spread of drought conditions across the area, as these trends continued.

The people of Akole Taluka mainly belong to the Mahadeo Koli, a Scheduled Tribe; there are also families of Mahars, a Scheduled Caste. Scheduled Tribes account for about 8 per cent of the population of India, occupying the lowest rung of the social ladder, socially, econom-ically and politically and living in conditions of extreme poverty, with limited access to education, healthcare and financial support, and to social services in general. The villages of Ambevangan, Manhere and Titvi have a total population of 3,239 (January 1993). Of the 494 households in the villages, 89 per cent are Scheduled Tribes, while 5 per cent are Scheduled Castes.

In early decades of this century, much of the taluka was still covered by extensive forest. Commercial logging by outsiders gradually stripped the area of vegetation, setting the scene for widespread erosion of soils by sheet flow and channelised run-off during the monsoon. The elderly among the people of Akole Taluka recall that their fathers and grandfathers spoke of a dense forest cover, as well as an abundance of surface water in the area. As a result, villagers recognise a connection between deforestation, increasing degradation of the land and reduced availability of water.

At the start of the project, the people were destitute, farming at a subsistence level of production. Only 3 per cent of households were without land. Among the other villagers, the average amount of land held per household was 4.3 ha. Nearly 31 per cent of families held between 2 and 4 ha of land; 10 per cent held more than 8 ha. In the *kharif* growing season (June to September), the main crops are rice, groundnuts, *ragi* and local grass; wheat and *gram* are grown during the *rabi* season (October to January). After the monsoon, lack of water forced many farmers to work away from home as unskilled labourers. Village women and their older children bore the hardships of locating and carrying water for domestic use. Many villagers had health problems (gastrointestinal disorders, skin diseases), related to water shortages.

3. Solutions

3.1 Needs assessment

Contact between the tribal and rural people of Akole Taluka and BAIF began in 1989 with initiation of the *Wadi* (Orchard) Programme. BAIF helped tribal families grow and cultivate fruit and other trees on plots of wasteland. Until then, people had experienced relatively little contact with outsiders. It was intended that families would become self-reliant over a period of several years, as a result of the sale of orchard produce. When the joint project with UW Earth Sciences started, this programme involved over 1,200 out of about 2,000 families in 14 villages.

BAIF conducted separate, rapid rural appraisals (RRAs) of water use and health in Akole Taluka in 1990, to assess needs as a basis for project activities. These RRAs were conducted by multidisciplinary teams, and involved discussions with individuals and groups of villagers. A major, participatory component was added over a period of several months, as villagers volunteered information to the BAIF field officers who lived and worked among them. Selected findings with particular relevance for this account are:

1. Villages commonly had two community-dug wells and several (on average, six) private dug wells. They were shallow, up to 10 m deep, and located within 500 m of the nearest dwellings, typically without parapets and lined with stone or cement.
2. Every village had at least one India Mark II Handpump with aprons, but as a rule the drainage system was inadequate. The wells were dry or damaged. No one in any of the villages had the technical knowledge necessary for maintenance of handpumps.
3. Some mountain streams maintained flow until January or February. A few natural springs discharged throughout the year. These water sources were used for domestic purposes. The Pravara river was a reliable source of water throughout the year, but was seldom used, because of the great distances involved.
4. These water sources were used for all domestic and livestock needs. The water – generally of poor quality – was used in close proximity to the source, collected and stored in brass and aluminium vessels. It was drawn by means of a glass and strained through cloth when seen to be turbid.
5. Malnutrition, vitamin deficiency and anaemia were common. Diarrhoea affected all age groups. Cases of typhoid and viral hepatitis were reported. Pneumonia and bronchitis were associated with

the monsoon. Acute skin infections were widespread in the summer months. Tuberculosis, leprosy and malaria were also common.

Two IDRC officials visited Akole Taluka in early 1991. They viewed the prospect of a water resource management project in such a dry area as daunting in the extreme. However, they commented on the presence of scattered trees, which clearly took water from an underground source. On that basis, they judged that the project had a chance of success.

3.2 Project life cycle

Projects can be divided into several overlapping stages, each characterised by particular outputs and evolving logically from the stage before. These are listed below, with special reference to contributions made by local people. The listing is approximately chronological, marking the first appearance in the project life cycle of a particular set of activities and related outputs.

3.2.1 The initial contact

The initial contact between the intending project partners began in summer 1991, with a visit to Akole Taluka by the UW project leader and members of the BAIF project team to assess the feasibility of the proposed activities. Discussions took place in village meeting halls and schools, and involved considerable formality in accordance with local approaches to communication. At the request of the outsiders, women attended the meetings and expressed their points of view. This deviation from custom was accepted by the villagers as a necessity, in view of the water-related hardships, traditionally borne by women. This stage in the project continued into the first field season, during the summer of 1992. The villagers repeatedly voiced their strong need for water to satisfy domestic requirements, regardless of quality. Field observations indicated that to provide a more accessible water supply for domestic use, the project would need to concentrate on reducing the volume of monsoon rain leaving the area as surface runoff. It was agreed that project activities would focus on watersheds around the villages of Ambevangan, Manhere and Titvi.

3.2.2 Sharing of information

Sharing of information by villagers and outsiders generally took place in farmers' fields. In the first field season, during the collection of hydrologic and hydrogeologic data at a reconnaissance level by members of the BAIF and UW project teams, farmers showed great curiosity about the sampling of water and solid earth materials by the outsiders and were eager to help in any way. They frequently insisted upon having a role in the sampling procedures. The villagers were fascinated by the links to earth satellites in orbit, employed in the global positioning system, used to provide precise map locations for sampling stations. They demonstrated a good understanding of explanations offered, in relation to the project purpose. The villagers generously provided insights into their own knowledge systems and religious beliefs. They took members of the project teams to sites of spring discharge that had ceased to yield water after the monsoon. They also talked about botanical indicators of ground water. This indigenous technical knowledge contributed greatly to the success of the project.

3.2.3 Selection of appropriate technologies

Selection of appropriate technologies required all parties to have a clear understanding of the practical steps needed to achieve the revised project purpose. Project research played an important role in selection, since it integrated field and laboratory studies of soils and rocks; onsite, hydrologic and hydrogeologic investigations; and indigenous knowledge of terrain features and botanical indicators of shallow ground water. The technologies deemed appropriate were mostly slope modifications, intended to reduce the velocity of monsoon surface runoff and to promote infiltration into the soil. Accordingly, project planning included expansion of the existing systems of terraces on hill slopes and a variety of water spreading and harvesting techniques. Many of these techniques are also effective in trapping soil eroded from higher elevations. Each village appointed its own watershed committee, usually by acclaim: these greatly facilitated the exchange of information necessary for implementation of the plans. The excellent working relations, established by the BAIF field assistants in their daily contact with the farmers, contributed to the trust that was essential for the introduction of new ideas.

3.2.4 Vocational training

People received vocational training, with an orientation towards making them self-reliant in the implementation of appropriate techniques for water spreading and harvesting, as well as selected methods of soil conservation. Women and men from each village were trained in a team approach to excavation of trenches and bunds (soil ridges), parallel to contours of elevation, using an A-frame. Training was given in the use of ferrocement and masonry in structures for water and soil

conservation, as well as in the construction of gabions (barriers, composed of rocks, bound together by chicken wire). In addition, villagers
were trained to install roof water harvesting systems, maintain handpumps, construct checkdams and the cleaning of dug wells. BAIF field
assistants and programme coordinators provided the training, and themselves received training in the field measurement of selected properties
of soil and ground water and the siting of water wells from members of
the BAIF and UW project teams. Much of this training took place during
the collection of research data. Two tribal families, in Manhere and Titvi
respectively, recorded daily rainfall and temperature data.

3.2.5   The development of demonstration sites

The villagers put their new skills to work in the development of
demonstration sites for implementation of water spreading and harvesting techniques. This work was coordinated by the village watershed
committees, in consultation with the farmers who owned the land, and
also by the BAIF field assistants, who assisted in the construction. The
existing system of terraces on the hillsides was also extended into areas
of wasteland, designated for agricultural development. Modifications of
traditional techniques of water and soil conservation included: gully
plugs and stone bunds, contour trenches and bunds, recharge pits and
ponds, farm ponds, dug wells and spring developments, masonry
checkdams, various micro-irrigation techniques and establishment of a
vegetation cover. Innovative techniques included: gabions, ferrocement gabion structures, artificial recharge near traces of fractures in the
bedrock, roof-water harvesting, use of impermeable soil to direct surface runoff into infiltration trenches, and an underground barrier to
ground water flow. The locations of demonstration sites were determined to a large extent by indigenous knowledge and analysis of
hydrologic and hydrogeologic field data. Careless disposal of body
wastes at outdoor locations by villagers was a constraint on storage of
water at low elevations.

3.2.6   Operation and use

Operation and use of the water spreading and harvesting technologies
have tended to confirm their appropriateness. During May 1997, more
than a year after the end of the project term, the people of all three villages had easy access to water in deepened, dug wells; in reservoirs on
the upslope sides of checkdams, trapping water from springs and seepages in the channels of ephemeral streams; and in tanks, receiving the
discharge of springs, maintained in part by artificial recharge. The

domestic water needs of the villagers are satisfied. Soil conservation measures implemented in support of these technologies were effective in arresting the downslope transportation of soil from hillsides and in facilitating infiltration of surface runoff. This stage also provided possible opportunities for individuals, involved in design, construction and water use at demonstration sites to learn from their mistakes. For example, in a three-level gravity-flow system near Manhere, the foundation of the bottom checkdam cracked, possibly as a result of vibrations set up by a nearby pump intended to supply irrigation water to the fields below, and the water of the associated reservoir escaped downslope. Villagers have gained an awareness of the importance of fractures in the underlying basalts as possible avenues of escape for waters impounded behind check dams, and they have deepened some reservoirs as a prelude to the grouting of fractures.

3.2.7 Replication

Replication of appropriate technologies employed in Akole Taluka, was directly related to their successful use and to the influence of individuals who were seen to have benefited as a result. People from the partner villages and adjacent areas visited the demonstration sites during all stages of development and in times of operation and use. Team members, BAIF field personnel, and local farmers answered their questions about design and use of the new technologies. The technologies employed were seen to be relatively cheap, small-scale and easily replicated. Villagers who had achieved success in the use of project technologies were taken as role models. Information about early project successes spread by word of mouth throughout the three villages and into neighbouring areas. BAIF was able to play a direct role in the transfer of project technologies across a wider area of Akole Taluka, thanks to the support of the National Bank for Agriculture and Rural Development (NABARD), under the Indo-German Watershed Development Programme.

3.2.8 Ownership of the introduced technologies

Villagers showed clear signs of preparedness to assume ownership of the introduced technologies, along with related responsibilities for monitoring. Farmers recognised that the breaks in slope, introduced as extensions of the existing systems of terraces and smaller-scale configurations of ridges and excavations, had the cumulative effect of increasing soil moisture and reducing erosion. They also realised that these effects are beneficial for agricultural production. Previously, farmers planted on the margins of terraces, which led to deterioration of the terraces and marginal bunds. Now they show concern about the maintenance of terraces and other conservation measures. Villagers have also seen that diversion of monsoon waters below ground level has increased the yields of nearby wells. Farmers have also seen the need to protect drinking water and reduce health risks to family members, and have requested training in the cleaning of dug wells. All of these initiatives represent investments in the future and a major change in the management of soil and water by the villagers.

3.2.9   *Dissolution of the partnership*

Dissolution of the partnership took place at the end of the project term. Unspent funds were reallocated for an assessment of project sustainability, carried out in the period April–May 1997. The management strategy for acquiring a year-round water supply was successful. The demonstration sites were effective in satisfying the water needs of the villagers. The problem of effective use of water resources remained, but at a reduced level. Villagers still washed clothes, utensils and other items at their water sources, but they showed an interest in hygiene and sanitation on a scale not previously witnessed locally: increasing numbers of pit latrines were in use in each of the villages.

4. Notes on project research

The indigenous knowledge of villagers, especially concerning the relationship of terrain to ground water discharge and biological indicators of shallow ground water, was an important starting point. Such information finds parallels in ancient Indian texts on science, such as the sixth-century *Brahat Samhita* of Varaha Mihira (Tagare 1992). For example, the tree *Ficus glomerata*, known locally as umbar, grows in close proximity to springs and seepages and at sites where the water table is shallow for extended periods. This tree – almost certainly the species noticed by the IDRC officials, who visited the area in 1991 – is revered locally (see also Feldhaus 1995: 112, n. 23).

Villagers were closely involved in considering baseline data and research results in order to choose appropriate technologies for possible use in the project area. Public meetings, following a traditional format, and onsite consultations with farmers and members of the local watershed committee provided the venues for these discussions. Many of the techniques of water spreading and harvesting discussed are known from dryland regions in other parts of India and elsewhere. Some of the catchment area modifications employed are related to techniques used more than 4,000 years ago by the Nabotean culture and its predecessors in the Negev desert (Nessler 1980).

High-technology applications included: imagery and location data from earth satellites in geographic information systems for the computer generation of maps and precise location in the field, by means of a global positioning system; chemical analyses of rock samples and chemical and isotopic analyses of water samples in the laboratory; and measurement of depths to the water table, the electrical conductivity of water, the hydraulic conductivity of soil and the magnitude of radon emissions near fracture traces. Research data were integrated as a basis for intermediate technology solutions to problems in water resource management and soil conservation. These solutions had to be compatible with existing local practice, readily understood, and easily implemented by the people.

The water resource management strategy developed employs demonstration sites for a wide range of intermediate technology approaches to water conservation. To a large extent, it builds upon the indigenous knowledge of the tribal rural people of Akole Taluka:

1. Techniques, referable mainly to the surface circuit of the hydrologic cycle, use barriers (contour bunds, checkdams, gabions) and shallow excavations (contour trenches, farm ponds, bedrock excavations), as well as natural and artificial surfaces. They complement the soil conservation function of the terraces on hill slopes under agricultural cultivation, exploit the short-lived, channelised flow of ephemeral streams, and also take water from road surfaces and adjacent gutters and the tiled roofs of dwellings.
2. The subsurface circuit is harnessed by means of shallow excavations to enhance recharge (recharge pits and trenches) and, at other locations, to contain the discharge of ground water (spring development) and through improvements to existing dug wells and bore wells for aquifer development. Gully plugs, intended to serve as sediment traps, also capture surface runoff. Simple tanks and other structures were installed for the development of springs and seepages. Underground dams were also employed.

This division is to some extent artificial, in that some of the breaks in slope of the first category also facilitate the infiltration of ponded waters. Alternative techniques for the extraction of water from the atmosphere as evaporation waters (modified Mexican still, solar still) and dew ponds were presented for possible use by separate families and individuals, under conditions of extreme water shortage.

A detailed account of project research is beyond the scope of this paper; some research activities, however, made major contributions to the selection and siting of water spreading and harvesting initiatives:

1. Field-saturated hydraulic conductivities of field soils were estimated, using the constant-head well method (Elrick and Reynolds 1992), applied to measurements from a field permeameter, constructed by technical support staff at the University of Windsor. These were combined with field values of porosity, bulk density and soil moisture tension in assessments of aquifer potential. Surface infiltration tests were also carried out. The data confirmed the occurrence of a thin surface skin of clay and silt above more permeable soils and were employed in determinations of the siting and density of recharge pits and trenches. In addition, observations of rivulet discharge in paddy fields suggested that the weathered bedrock below the soil was capable of storing and releasing significant amounts of water.

2. A reconnaissance geological map was prepared, starting from the work of Subbarao and Hooper (1988) as a starting point. Straight-line ground features (lineaments), mapped on the basis of imagery from earth satellites, were overlain on the geological map. In the project area, lineaments commonly defined belts of vertical fractures, breaking the lateral continuity of the impermeable basalt lava and providing possible conduits for the migration of ground water. Existing dug wells, sited on the surface traces of fractures, were selected for deepening. Springs and seepages were also localised, where lineaments crossed hill slopes. The discharge of these springs was augmented by means of artificial recharge at higher elevations. The waters of springs and seepages, in the valleys of ephemeral streams that coincided with lineaments, were impounded on the upslope sides of checkdams.

3. Water samples collected from dug wells and surface locations in May 1992 and December 1993 were analysed for environmental isotopes (oxygen-18 and deuterium) at the University of Waterloo. These data, combined with measurements of electrical conductivity, made onsite, gave an approximate indication of the nature and location of ground water recharge. The well waters had isotopic signatures, similar to those of surface waters; they had undergone evaporation, consistent with a shallow mode of occurrence; and there was no basis for invoking a deep source of ground water. Therefore the recharge of well waters through the weathered

bedrock and fractures in the underlying lavas is probably referable to the most recent monsoon event.

The BAIF field officers were informed about the significance of the research results to the project purpose and passed this information on to the villagers.

5. Results

All 494 households in the area now have relatively easy access to water for domestic and agricultural uses. Up to 20 per cent of the households obtain water for domestic use from six developed springs. Roof water harvesting systems have been adopted in all three villages, in 26 households altogether. In part, the decision to adopt this approach was related to availability of water from other sources. About 73,000 m^3 of water are stored in the reservoirs behind 14 masonry checkdams and three ferrocement gabions. Water availability to the villagers has increased by about 750 litres per day per capita during the driest part of the year, at the end of the pre-monsoon season. In addition, artificial recharge has diverted possibly as much as 19 million m^3 of water each year into the soil.

The measures to reduce volume and velocity of surface runoff had the additional effect of greatly lowering soil erosion. Contour trenches seem to have prevented an annual soil loss in the order of 150,000 tonnes. A further 6,550 gully plugs, 75 dry-stone bunds and 75 gabions have trapped about 32,000 tonnes of soil altogether. The increased soil moisture permitted the villagers to harvest a second (winter) crop from 75 ha of land. About 300 ha of land were brought under cultivation during the rainy season. In 1995, the village watershed committees reported that 5–15 per cent of farmers took produce to sell at local markets.

Some 25 village youths were trained in the use of ferrocement and masonry in water and soil conservation measures. Between 10 and 15 people per village were trained to construct gabions. In the three villages, 6–10 men and women per village received training in team approaches to laying out trenches and bunds, parallel to contours of elevation above mean sea level. Considerably larger numbers of villagers received instruction on installation of roof water harvesting systems, maintenance of handpumps, construction of checkdams and use of artificial-recharge techniques. Villagers were also instructed in the cleaning of contaminated sediment from the bottoms of dug wells. This training was usually done at the request of villagers, who demonstrated management capability and leadership potential.

Increased employment opportunities now exist in and around the villages, so that farmers no longer feel under pressure to work away from home. Villagers who still elect to work away from home have improved employment prospects for better pay, as a result of the skills developed from vocational training. Individuals and village cooperatives, including women's groups, are planning new, agricultural enterprises, such as the acquisition of mechanised milling facilities and the introduction of new crops, including tomatoes. Some villagers plan expanded uses for excess water, such as in irrigation schemes. Village women no longer must spend the greater part of each day searching for and carrying water. The time saved is spent working in the fields. As more land goes into agricultural use, livestock quality is improving and livestock numbers are decreasing.

Health problems associated with water shortage now show reduced incidence. Seeing a connection between deteriorating water quality, sickness among farm workers, lost time in the fields, declining productivity and lower financial returns, villagers have sought instruction on the cleaning of dug wells. Women strain water from dug wells through several layers of *sari* material at the well site. Basic instruction in hygiene was provided to villagers in individual households, in the village schools and at public meetings. Women were receptive to the idea that they take on the role of environmental managers, but the response was always that 'it will take time'. The use of pit latrines is steadily increasing, and domestic wastewater is used for cultivation of kitchen gardens. Sanitary conditions in the houses are improving and about 200 families have gained health and nutrition benefits from kitchen garden produce.

Local people show greatly improved morale, evidenced by more outgoing attitudes, increased attention to personal appearance and better upkeep of houses. At public meetings, teenagers said that they were optimistic about the prospect of life in a village setting and presented arguments against moving to a town. Several new housing starts in the villages indicate a confidence in the future among people who formerly placed greatest value on short-term gains. Indeed, villagers' adoption of project technologies to such good effect is itself evidence of a fundamental change in their way of thinking to pursue long-term objectives. Villagers now want to participate in group activities for the benefit of the community, indicated by a rising number of individuals expressing interest in joining a village watershed committee to advise others on the management of water resources.

In the past, villagers understood environmental processes affecting their lives well, but on a scale that was limited to their immediate surroundings. News about the demonstration sites of the project spread across Akole Taluka by word of mouth, especially during the early stages. With the passage of time, people saw that similar water and soil conservation initiatives were successful across a wider area and, as a result, began to perceive their relationship to the environment on a much larger scale. This expanded perception promoted replication of the appropriate technologies employed in the project and contributed to wider adoption of them in neighbouring areas.

The water resource management strategy, developed in the project, is readily transferable to the rest of the Deccan Trap region, on the basis of similarities in hydrogeologic environment and ground water regime. The urgency of the need for a management strategy applicable to water resources across the nation was recognised well before the start of the project in the National Water Policy (1987), formulated by the Ministry of Water Resources, of the Government of India. More recently, successful application of project results coincides with a statewide programme to provide drinking water to villages, initiated by the Government of Maharashtra.

6. Lessons learned

Several lessons were learned during the project and were helpful to the project teams in their continued interaction with the people. Other lessons came later, through critical appraisal of the various directions, taken by project activities. These lessons, spanning the entire project term and the year following its termination, include:

1. Poor tribal, rural people were initially suspicious of the motives of outsiders. They were reluctant to go to their fields with strangers, for fear of losing their property to the newcomers. At the start, villagers did not welcome the prospect of having farm ponds and other excavations on their land, because they felt that the area of cultivation or grazing would be reduced. However, they responded positively to suggestions that knowledge of water and solid Earth materials might be exchanged.
2. Interest in and respect for people's religious beliefs paid dividends for project activities, in that they revealed valuable insights into the relationship between revered trees and underground water. The recognition of botanical indicators of shallow ground water became a standard, initial step in the selection of sites for deepening of dug wells and blast-holes and development of springs and seepages.

3. Because BAIF's field officers lived in the villages and shared the daily hardships of the people in all seasons of the year, they won the trust of the villagers and developed effective working relations. Their close proximity on a day-to-day basis also helped the exchange of information and added useful participatory elements to data from the initial RRA. Daily discussions provided a basis for fine-tuning the project activities to meet local needs better.

4. Villagers who were seen to have achieved success in water and soil conservation, as well as improved crop production, as a result of adopting the techniques of the project, served as role models within their villages and for the people of neighbouring areas. These role models played an important part in contributing to wider use of the technologies within the collaborating villages and adjacent areas.

5. The hydrogeologic environment of the project area initially appeared to be unpromising for successful conjunctive use of water resources. A major obstacle was the poor aquifer potential of the lava sequence. However, a research focus on fracture-trace analysis provided a systematic rationale for locating springs and seepages for development. This was consistent with the villagers' practice of siting dug wells, where the lavas are transected by vertical fractures.

6. The beneficial changes affecting the people, involved in the project, are best described by IDRC's dictum 'empowerment through knowledge'. They are in part a consequence of the wide-ranging influence of increased availability of water on different factors, contributing to the quality of life: nutrition and health, material standard of living, and considerations of personal motivation.

7. The successes of this project have created an enabling environment for similar activities in the surrounding area. Other villages are forming partnerships with outside agencies to develop the scarce water resources more efficiently by introducing the same technologies. Indeed, since the initiation of the IDRC-funded project, BAIF itself has used these technologies successfully with other groups of villages as partners over a wider area of Akole Taluka.

8. The time-frame for evaluation of sustainability is of vital importance to the success of a project on water resource management. The project was allowed to run for one year more than its original three years, with no additional funding, and its sustainability was assessed more than a year after termination. This was made possible through frugality in the use of funds and also flexibility on the part of the funding agency.

7. Concluding remarks

The project results were considered to be sustainable on the basis of the following indicators:

1. The tribal and rural people of Ambevangan, Manhere and Titvi helped design the project and contributed to the activities at every stage. Their comments, related to the evaluation of project outputs, were communicated on a regular basis, so that additional action could be taken, when needed.
2. People received vocational training in support of planning, acquisition and use of the water spreading, water harvesting and soil conservation measures, introduced during the project. This knowledge has provided them with the technical capability to sustain the project results.
3. The villagers have assumed ownership and responsibility for sustaining the project. They see that their water needs are satisfied, agricultural production has undergone improvement, and they can withstand some fluctuation in the amount of monsoon rain, as a result of new land brought under production.
4. Though labour-intensive, the appropriate technologies for water and soil conservation are cheap, small-scale and easily replicated. With continued improvements to agricultural production, the villagers will have all the sustaining resources needed to operate and maintain the project results.

Effective use of the technologies is another important component of project success. Most of the systems introduced show a high degree of reliability. The quality of water at the source remains good for the most part and maintenance of the water spreading and harvesting technologies presents few problems. Hygienic use of the facilities shows a steady improvement, as the villagers develop environmental consciousness.

Replicability of the technologies is evidenced by the readiness, with which they have been accepted throughout the project area and in neighbouring areas. This is in part related to the nature of the technology, which augments existing reciprocal relations with respect to natural processes.

The transfer of technologies from Akole Taluka over a much wider area was promoted through use of project results in the training of BAIF personnel in Pune and Akole Taluka, during May 1997. Twenty-four programme coordinators and field personnel, representing ongoing projects in five Indian states, took part in a week-long training course on water resource management. Villagers and BAIF field officers participated as instructors. Project results also formed the subject matter for a workshop on watershed management for programme personnel of BAIF-GRISERV, held at Sanosara, near Bhavnagar, Gujarat, in June, 1998. The intention is that dissemination of project results in this way will hasten their adoption over a wide area, extending beyond the Deccan Trap.

Note

1. The authors – the Canadian (FS) and Indian (GS) project leaders – gratefully acknowledge the guidance of Manibhai Desai, the late founder and first President of BAIF. They also appreciate the help of his successor, Narayan Hegde. The International Development Research Centre, Ottawa, funded the project.

Part IV

Learning from Success: Supportive National Policies and Local Initiatives

10
Creating New Knowledge for Soil and Water Conservation in Bolivia

Anna Lawrence

1. Introduction

Attitudes to soil erosion have changed over the last 60 years.[1] It is no longer approached as an isolated technical problem, but one closely linked to productivity maintenance and the elements of interacting land use systems within a watershed, as well as to social and political issues (Blaikie 1985; Young 1989; Stocking 1996). Research has shown that the 'problem' identified by outsiders does not always exist, or is addressed in a different way by farmers (Leach and Mearns 1996; Scoones 1997), and that farmers' own knowledge of soil fertility management was neglected in many earlier extension campaigns (Kerr and Sanghi 1992; Pretty 1995). Consequently, recommendations by scientists for soil and water conservation (SWC) have not always been widely adopted, because the problem addressed is not perceived by farmers, the solution is too expensive, social factors intervene or farmers already have superior practices (Fujisaka 1994).

In these circumstances the old-established models of technology transfer whereby researchers identify solutions and extension agents pass them on to farmers are inadequate (e.g. Long and Villareal 1994; Okali *et al.* 1994). If scientific or other 'external' knowledge is to contribute to increased rural sustainability, it must work in different ways. Long and Villareal (1994) suggest that replacing the notion of 'transfer of technology' with that of information systems and communication theory is not even enough; instead, they argue, we must understand that knowledge is *transformed*, not transferred, when it crosses social interfaces.

In this chapter I discuss a 'participatory soil conservation' project in a part of Bolivia where soil erosion is perceived to be disastrous by researchers and development workers, yet neither they nor farmers had hitherto addressed this problem. I explain why a participatory approach was chosen, suggesting that the reasons were pragmatic rather than informed by theory. In analysing the results, a more theoretical critique is seen to be relevant in helping to understand the effect of the dynamic knowledge interface that was facilitated by the project.

In the temperate valleys of Santa Cruz department, Bolivia, a long history of settlement has brought land degradation through intensified use of sloping land in a semi-arid climate. Only 14 per cent of the area is classified as suitable for arable crops but much more is cultivated (Davies 1994). The valleys form a heterogeneous zone, being in the area of transition between the Andes and the dry Chaco, with altitudes ranging from 700 m to almost 3,000 m above sea level, and annual precipitation from 350 mm to almost 2,000 mm per annum.

The three valley provinces of Vallegrande, Caballeros and Florida have a cultural and ecological unity, but this is changing as migrations occur along the trunk road to Cochabamba. One community from each province was selected to participate in the project, and their differences serve to highlight the complex context in which natural resource management is evolving. Apart from ethnic and agroecological differences, they represent three stages of rural population processes: agricultural expansion (Pozuelos), stability (Chacopata) and decline (Los Pinos). All households in these communities are involved in farming, notably in producing crops for market, but many supplement their income through seasonal agricultural wage labour in lowland Bolivia. Temporary migration often turns into permanent emigration as young people leave to study and then find work in the city of Santa Cruz.

Chacopata, the oldest of the three, was established about 200 years ago, at about 2,200 m above sea level in the semi-arid *valle alto* (high valley) zone. It follows the Vallegrande system of maize production on the lower slopes, with extensive cattle production on less accessible land; recently, farmers have also begun fruit production. Ethnically the vallegrandinos are *mestizos*, of mixed Spanish and lowland indigenous origin. Los Pinos, near the border with the highland Cochabamba department, was settled by ethnically distinct *quechua* families who left their homes in Cochabamba about 70 years ago due to pressures on the land. The highest of the three communities, at 2,400 m above sea level, near the cloudforest, it is cooler and more humid than the other two. Pozuelos is the lowest of the three, at 1,400 m. The community was established at the beginning of the twentieth century by out-migrants from Vallegrande. On the highway to Santa Cruz, it attracts more recent (*quechua*) migrants from Cochabamba especially since the hillside forest has been opened up by roads built since 1993. Old and new settlers alike have adopted a system of rapid clearance, production of chilli peppers and beans on these slopes. Of the three communities, only Pozuelos has irrigated land, on riverside areas where vegetables are grown.

In all three communities livestock are significant in the farming system. Cattle and sheep are grazed on extensive ranges often far from the community, but are brought in during the dry season to graze the crop residues. While men are generally more involved in decisions related to annual cropping, women tend to take care of small livestock including sheep, and to increase their responsibility for the cattle as male labour becomes less available (Painter 1995). Overgrazing and deforestation are considered to have been problematic for at least 200 years (Vargas *et al.* 1993), but contemporary observations provide no direct evidence of this.

The more recent move uphill of Pozuelos farmers is reflected in several differences between its farming system and those of Chacopata and Los Pinos. The cultivation cycle is longer in Chacopata and Los Pinos than in Pozuelos where there is still new forest to open up, soil is washed away more quickly, and weed growth is faster. Agrochemicals are used in Chacopata and Los Pinos, but not on the slopes of Pozuelos where soil fertility is higher, and pests not yet a problem.

2. Methodological approach: creating a 'knowledge interface'

The institutional linchpin of the project is the state-funded Tropical Agricultural Research Centre (Centro de Investigación Agrícola Tropical, or CIAT, not to be confused with the CGIAR centre based in Colombia). After decades of commodity-focused research, CIAT has adopted a farming systems approach, and simultaneously has gained experience in a range of participatory methods (participatory rural appraisal (PRA) and on-farm 'validation' trials). These approaches have usually been introduced through donor-funded projects, and in an institution of more than 100 technical staff the experience of each project is often isolated from that of others. The project discussed here is also donor-funded, and the 'participatory research' approach was a further addition to the gamut of 'participatory' experiences of CIAT.

Contrasting notions of participation were only vaguely elucidated at the start of the project, and consequently had to be analysed later. The participatory research approach explicitly contrasts with the normal model of technology transfer used by CIAT, whereby it interacts with a large network of NGOs which are termed 'intermediate users', intended to adopt CIAT's research results and pass them to farmers. In the temperate valleys there are two constraints to this model: relatively few NGOs are working in agricultural issues, and the territory is so diverse that developing technology in a centralised way is inappropriate for diffusion to a large number of farmers. The existing links between CIAT and NGOs in the area provide a strong precedent for following a new model of technology development, enabling multi-directional learning.

Of the few NGOs operating in the temperate valleys, CIAT approached four to collaborate in the project, two focusing on community organisation, together with one with a background in biodiversity conservation, and one organic agricultural cooperative. Soil conservation was already a priority activity of the latter two, but not in the higher semi-arid area where this project focused. The recent Law of Popular Participation (1996) has allocated budgets to elected community organisations (Organizaciones Territoriales de Base, or OTBs), and most NGOs are currently preoccupied with facilitating community decision-making to plan socio-economic development and (more rarely) natural resource management.

It is helpful to conceptualise the project methodology in terms of a 'knowledge in action' framework (Blaikie *et al.* 1997), which adopts the concept of a 'knowledge interface', recognising the evolution of rural people's knowledge (RPK) and the dynamic interaction with 'outside knowledge'. This project aimed to create the conditions for a creative interface, not only by *identifying* knowledge, but also by *bringing together* the holders of different types of knowledge. This was the most efficient approach under the conditions of existing knowledge. CIAT had for some time recognised a need to work in the temperate valleys, in response to the perception of disastrous soil erosion, but had not tested any soil conservation technologies in the zone. No documentation of farmers' knowledge of soils or the environment, nor of their soil management practices was available. While CIAT, farmers and NGOs might be assumed to have relevant knowledge, it did not add up to knowledge applied to the perceived problem of soil erosion. This approach, born of development practice, bears analysis within the theoretical discourse currently surrounding information and technology transfer, as I will discuss later.

The methodology consisted of two iterative stages: the first, to explore relevant knowledge and practices in the temperate valleys (among farmers and institutions), and the second to facilitate the use

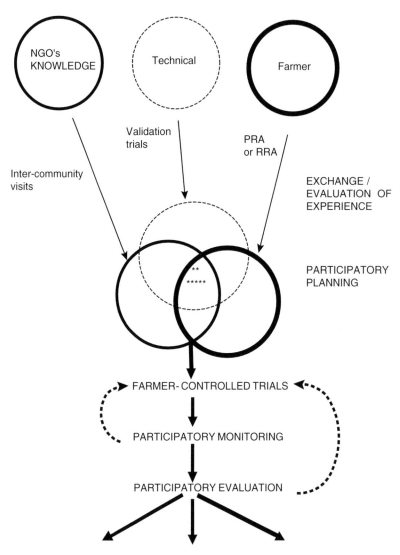

Figure 10.1 Stages of information flow in participatory research

of this knowledge by a range of farmers, in experiments of their own design on their farms. The process was conceptualised as shown in Figure 10.1, envisaging information flow between the various steps of the project. This plan is based on a communication process of which the pivot is 'information exchange workshops' involving stakeholders possessing different knowledge.

The 'knowledge exploration' stage was the starting point for a cyclical process. CIAT conducted a quick appraisal of institutions working in the area, and of documentation of any available technologies for soil conservation. These initial activities also served as introductions, helping to include NGOs and farmers in the process and encouraging them to contribute to the 'information exchange workshop'.

A short and focused form of PRA was then used to explore relevant rural knowledge. One tool which was developed for this was a matrix, prepared by village groups who brainstormed practices used by people in the community for reducing soil erosion. For each practice, the group then listed where they had learnt it from, listed reasons why some people did not do it, and ranked the practices to gain an idea of how many farmers were actually using them. This and community mapping of natural resources were group activities which raised interest among community members and convinced them that the researchers were interested in their knowledge, rather than expecting to tell them what to do.

3. Understanding rural people's knowledge

Researchers and development workers in the temperate valleys often claim that 'the farmers don't know about soil conservation', and many farmers themselves report that they do not do anything to prevent soil erosion. By exploring knowledge and practices however, the study showed that an apparent lack of knowledge is linked to differences in perception of the problem, and terms in which solutions are framed.

3.1 Perceptions of soil-related problems

Despite decades during which outsiders had defined the problem of the valleys as 'soil erosion', this did not fit with the perception of farmers. Their day-to-day livelihood priorities, and the dynamics of a changing system, contribute to their own perceptions of different problems. This was evident in the differences between the views and attitudes of people in each of the three communities, taking into account their different histories.

The first difference of perception is in terms of the way the 'problem' manifests itself. Little traditional knowledge specifically addresses soil erosion, but all the respondents observed that the quality of their land has declined over the time that it has been farmed. Fertility decline is attributed to continued extraction of crops, declining organic inputs, continued burning and the increased use of chemical inputs, and showed understanding of the value of soil organic matter in maintaining fertility and humidity. Experience from other countries suggests that farmers are commonly more concerned about fertility loss than about erosion (e.g. Gardner 1997; Leach *et al.* 1997); while the former is visible as declining productivity, the latter may not be highly visible, or may not have a strong impact on their lives. The decline of fertility is most noted in the older communities, Chacopata and Los Pinos, and here farmers have taken most action to protect the land. Ironically erosion *is* perceived in Pozuelos in the wake of deforestation over the last five years, but because there is still more land for them to open up to agriculture, the level of concern is much lower than in the other communities.

Above all, the three communities have in common the fact that they prioritise *water* loss as the principal problem related to natural resource management. Many of the consequences which outsiders perceive as problematic could equally be addressed through water conservation.

A second difference relates to perceptions of causality. Farmers link declining soil quality with deforestation and continuous cropping. By contrast, the cause that is cited most strongly by outside observers – overgrazing – is not widely recognised in the communities. In some cases farmers denied that grazing was linked to land degradation at all. With the growth of markets and improvement of farmers' access to market, more land has been converted to annual crops in the last two decades, and livestock numbers have declined. Furthermore, with reduced burning, the vegetation may well be recovering from an earlier, more degraded state. The 'problem' of overgrazing is more complex than has been indicated by researchers, and is related to the different roles of cattle and sheep, and the shifting balance between them. Cattle roam over distant ranges, causing little visible erosion but also providing little manure except in the dry season when they are brought in to browse the stubble. By contrast, sheep (which are decreasing relative to cattle) are kept closer to home (competing more with land used for crops), are corralled every night, and provide manure for use on the crops.

This closer intimacy with cause and effect, and experience of continuing change, affects perceptions of sustainability. Farmers in Chacopata do not consider their farming systems to be unsustainable, partly because they are continually adapting to new markets and information. The increasing popularity of fruit trees in all three communities is a prominent example of this. In Los Pinos and Pozuelos, there is more implicit anxiety shown about continuity of farming systems because of out-migration; many of the younger generation have left to find education and work in Santa Cruz, or as far afield as Argentina. A particularly strong feeling in Pozuelos is that in 20 or 30 years' time no one will be left to farm in the community, and that there is therefore little point in worrying much about soil conservation. Thus farmers may be doing more than is apparent to prevent land degradation, by addressing a different aspect from that identified by outsiders.

Furthermore, the communities differed in their responses to the project's initial approach. Although soil erosion is not conceived as the central problem in Chacopata, most enthusiasm about the project was expressed there, while in Pozuelos the idea of addressing the more visible soil erosion was greeted with some apathy. These initial responses have been modified through three years of project interaction, but they are reflected in the practices and innovations which farmers were already using before the project began.

3.2   Technologies and management strategies

Table 10.1 summarises the different soil management practices encountered in the communities, showing how technologies have evolved over the last few decades. The traditional methods relate largely to the conservation of fertility and to the management of water, consistent with farmers' perceptions of the key constraints to agriculture. More recent innovations are directed towards control of soil erosion, and are currently practised by a few individuals in the older villages. The most common reason given for developing these new practices, however, is the conservation of soil organic matter, showing the importance of links between humidity and organic matter, as perceived by farmers. Reduced burning and leaving fallow or crop residues in contour lines are more widely practised in Chacopata and Los Pinos than in Pozuelos, where slope clearance is very recent.

In some cases, community action has been taken to control widely perceived problems. Of more universal concern than soil erosion is the loss of water sources and soil humidity, and many communities have protected their water sources and legislated to reduce burning of fallow. In Chacopata a few farmers have addressed overgrazing by deciding to delimit individual property within group-owned grazing land, which allows them to fence and manage it more sustainably.

3.3   Differences in knowledge and its application

Rural people's knowledge is differentiated according to both personal and contextual factors (Fairhead and Leach 1994; Long and Villareal

Table 10.1   Rural people's practices for SWC in the temperate valleys

Traditional (practised by parents of the current generation)	Innovations (newly developed by individuals)	Outside practices (under formal trial in the area)
• Fertility and water management • Drainage trenches • Digging in weeds • Leaving weeds under fruit trees • Grazing livestock on stubble • Expanding uphill edge of fields	• Could be interpreted as relating to erosion Control, but farmers explain them as intended to maintain levels of soil organic matter • Not burning fallow or crop residues • Leaving trunks/roots in the soil • Leaving contour strips • Contour-planting fruit	• Contour hedgerows • Cover crops

1994), and this is demonstrated in the way that people's experience shapes their attitudes to land conservation. In the temperate valleys of Santa Cruz, farmers' experience, knowledge and attitudes towards the use of that knowledge are related to the age of the community, migration patterns, tenure, gender roles, organisation and farming systems. The longest settled communities, where farmers have observed changes over decades or centuries, have shown most concern about land degradation (whether loss of soil, fertility or water) and have taken most action to address the problem. Concomitant with this history is the fact that these communities have limited land; only in Pozuelos do households still have forestland they can expand into. Migration is related to this; farmers who have recent experience of migration expect to be able to do so again, and therefore do not concern themselves as much with long-term decline in the area they currently inhabit.

Tenure differences within Pozuelos, and between the three communities, highlight the importance of landownership systems. Many hillside farmers in Pozuelos are tenants with a contract to clear the forest and farm for three years, not one of whom is interested in adopting contour hedgerows, whereas owner-tillers show some interest. In much of Vallegrande, the undivided land system – whereby a group of families owns jointly a pasture area – blocks any individual interest in conserving or improving the land. In a community near Chacopata, when the owners divided one such area, they reduced the cattle load dramatically, because they realised their piece of land could not support those numbers.

Men mentioned many of the new innovations for soil conservation, because they related to the cropped land. Women's knowledge is a distinct and important resource, but one that was initially undervalued by research staff. The project found differences between women's roles in different ethnic communities. In all the communities, the women tend to take care of the small livestock, and where the men are often absent, this responsibility often extends to cattle as well. In *quechua* households the women may travel far with the sheep or cattle and take an independent view of their management. Although women in the vallegrande communities also take care of small livestock, they perceive a more home-oriented role for themselves, focusing on family care and orchards near the home. This may be because the men in these communities migrate less for seasonal work, than do the *quechua* men. Nevertheless women in all the communities have detailed knowledge about the behaviour and management of livestock, but feel less able to make decisions about crops. Clearly, both components are essential to a holistic understanding of the farming system, particularly in an area where overgrazing is widely considered to be the underlying cause of the land degradation.

4. Outsiders' knowledge

Having elucidated the ways in which rural people were addressing issues of land degradation, the experience of institutional stakeholders completes the picture. CIAT had no prior experience of SWC practices suitable for the semi-arid hills, its research in the zone being restricted to the production of fruit trees and protection of pasture by fencing. Over the period of this project, CIAT introduced 'contour hedgerows' through on-farm trials in partnership with another project (Sims *et al.* 1999), but in many ways CIAT's role in this project was less as a source of knowledge, and more as a facilitator, bringing together those with relevant experience.

The valleys have for a long time been the preserve of NGOs and specific donor-funded projects, but in the semi-arid zone these have little or no experience of SWC. Others working at the more humid edge of the zone, nearer to the city, however, do have experience relevant to the project. The organic cooperative encourages the incorporation of crop residues into the soil, integrated pest management, and less erosive irrigation systems. Most valuable in terms of existing knowledge was that related to a FAO-funded programme (SEARPI) which had promoted contour hedgerows and cover crops to farmers in the catchment area of the Piray, a tributary of the Amazon which supplies Santa Cruz city. Although both the organic cooperative and the FAO programme work in the lower and more humid zone, they provide an important source of farmers' own experience, and these farmers provided the knowledge that most stimulated farmers from the drier zones, the project participants, to try out new ideas.

5. Creating new knowledge

5.1   The stakeholder forum

The project has gone through iterative cycles of information exchange and experimentation. The first cycle was the most ambitious, bringing together NGOs, FAO consultants and farmers to present experience and innovations, and then support farmers in planning trials. Later cycles were more local, and knowledge was shared between local farmers and CIAT. By this time, 'outside' knowledge had been adapted into the community, and there was less need to travel long distances to visit more experienced farmers.

Farmers travelled to the first workshop from communities in two provinces. One day was allocated for information exchange and one day for planning experiments. On the first day, farmers from each community presented summaries of their experience with SWC technologies, similar to those listed in Table 10.1. The PRA experience helped CIAT staff to guide the questioning afterwards, showing how ideas had originated with the farmers and spread to others. NGO representatives then made short presentations of technologies that they were promoting. In the afternoon, all the participants visited farms in the FAO-funded project, to discuss with the farmers their experience with contour hedgerows and barriers, and cover crops in fruit orchards. The field visits impressed the participants, many of whom collected seed from the barrier and cover crop plants, to test out at home. The workshop organisers had had doubts about the usefulness of visiting a warmer, more humid zone as a demonstration area, but this did not discourage participants, who during the second day included much of what they had seen from the field in their plans for trials in the coming season.

A participatory planning process was used on the second day and is summarised in Box 10.1. The day began with a plenary session, intended

---

Box 10.1   The process used to help farmers plan experiments during the participatory workshops

In group:
- Review of practices reported and seen yesterday
- Brainstorming of all problems linked to loss of productivity on farm
- Sorting of problems into a problem tree
- Linking of suggested solutions to problems

Individually:
- Draw plan of farm
- Indicate in different colour the suggested changes to be tested
- Present back to group, and discuss
- Prepare calendar of activities for implementation of trial

In group:
- Brainstorm list of indicators in response to question 'How will I know if my trial is giving me good results?'

---

to emphasise the relationships between cropping and livestock components of the farming system. Almost all the SWC work in the zone to date has focused on cropping systems, whereas many of the problems are assumed to originate with overgrazing and extensive cattle management practices. However, the problem tree method proved somewhat confusing to facilitators and farmers alike, and most of the proposed experiments still related to cultivated land. In addressing SWC, more work is still needed to relate these two components of the systems.

After a self-evaluation, the CIAT team proposed a second workshop with NGOs working in a third province. This gave CIAT and the NGOs the opportunity to build on the experience of the first workshop and propose improvements to the process. In particular, CIAT wanted to make more effort to involve women, find more positive ways to include those who lack confidence in reading and writing, and improve their use of the problem tree analysis. Their efforts were well rewarded, and 30 farmers attended the second workshop. The NGOs had been more closely involved in the organisation (and also showed more commitment to the proceedings) than in the first workshop.

In the second annual cycle, CIAT staff planned the workshops without external facilitation, and focused on the communities where trials had already been established. More women participated in these workshops, because it was not necessary to travel. Almost all the farmers who had participated the previous year were trying contour hedgerows, lines of trees or grasses planted along the contours. Some farmers wanted a single line of trees; others (mainly women) wanted to use the new varieties in pure pastures or woodlots. Initial reluctance among CIAT staff to support these modifications gave way, and by the second year technical staff made fewer interventions and were more appreciative of farmers' own results and explanations. CIAT focused more on facilitating communication between farmers. Nevertheless farmers' choice of technology was clearly influenced by the availability of free planting material, to which they contributed their own labour in setting up trials.

5.2   New knowledge applied?

Over three annual cycles of information exchange and experimentation, many farmers changed their farming practices, their perceptions of soil erosion and their opinions of introduced technologies. Yet there is no clear sense in which a technology has been 'adopted'; the response of farmers has been more complex, and other farmers have not changed their practices. To understand these responses we have to draw on the growing methodology of participatory monitoring and evaluation (Abbot and Guijt 1998). Through the use of participatory tools such as ranking and scoring, semi-structured interviews and recording forms (Lawrence *et al.* in press), farmers' own responses to their trials indicated their doubts and motives for trying out new ideas, and the ways in which new knowledge has evolved through its application. This broad approach to monitoring suggested that context affects not only farmers' existing knowledge and approach to the knowledge interface, but also their access to that interface and ability to put new knowledge into action.

Initial doubts about new hedgerows (that they obstructed the plough, and were not palatable to cattle) were gradually replaced with more positive observations, that the hedgerows were stopping the wind, retaining moisture, reducing disease, providing fuelwood, and above all, providing fodder. Farmers' attitudes to SWC have changed as they have tried out the new ideas. They do now believe it is possible to halt the flow of soil, in contrast to their earlier view that erosion was a 'natural' process, but their motives for starting were a mixture of curiosity, free inputs, enthusiasm of participants in other areas and expectation of enhanced fodder production. Their reasons for satisfaction, and for setting up nurseries to produce their own trees and grasses for future contours, are rather different from those envisaged either by themselves, or by CIAT.

Above all, paradoxically, CIAT wanted to work in the area because of the perceived problem of overgrazing, but despite a very open approach to technology development, has only succeeded in supporting change in crop management. Livestock management issues have not been addressed, for two reasons. First, CIAT staff had no interest in including women in the project, believing that they would just reflect their husbands' views; and second, CIAT is a technical research organisation, which shies away from the social challenges of changing the management of shared land. Women have been involved more as the project develops, and are more interested in fodder production. With time, the way that new knowledge has been used in rural communities may turn out to have more relevance for addressing livestock management, through a roundabout route not anticipated by project staff. Some farmers are finding that the hedgerows produce enough fodder of sufficient quality, to keep tethered cattle closer to home. One farmer has considered replacing his unproductive cows with fewer, improved stock. These kinds of transformations in livestock management have allowed increased population density to be accompanied by increased biomass and reduced soil erosion in other parts of the world (Tiffen *et al.* 1994; Turton *et al.* 1995).

More farmers participate in workshops and trials in the higher communities, where decades or centuries of farming have made them more conscious of land degradation. In Pozuelos there is no concern for environmental protection; the resources are newly available and represent an unexpected bonus, so they are not managed sustainably. Residents still enjoy the short-term economic benefits of renting their land or producing higher yields than on their lowlands, so interest in project activities is superficial.

But the explanation of differences is more complex than that. There were also differences in community organisation, and previous experience with agricultural extension. In Chacopata farmers have responded with enthusiasm to the project, because of established relations with CIAT and NGOs. In other communities, notably Los Pinos, it emerged that local families were controlling information flow and excluding others from participating in workshops. Knowledge creation and processes are affected by social and power relations, not only between farmers and outsiders, but also within the community. At an early stage, however, project staff realised that their preconceptions about *who* should be involved would have to be revised. For example, they assumed that the 'target group' in Pozuelos comprised those who lived on and farmed the upper slopes. But this would have excluded most farmers who live in the valley but cultivate a range of land types, those who manage cattle in the higher forest and landowners whose perceptions of soil erosion are developing rapidly. Working only with those perceived to be experiencing soil problems would have led to a misunderstanding of the changing ideas of the community, the conflicts within it and the potential for working together.

6. Learning from change

Part of the participatory evaluation approach is the use of self-evaluation as a tool for learning and institutional-strengthening. In this project it indicates four key points about the sustainability of the approach:

1. the CIAT project team has spontaneously taken up the model as a means of planning participatory research;
2. the project, through advocating a more radical participatory approach than those before it, triggered an internal review within CIAT to take stock of its accumulated experience with participatory methodologies;
3. the project has facilitated and worked with linkages with other organisations in the zone who may be in a stronger position than CIAT to continue using the approach; and
4. CIAT staff (beyond those directly involved in the project) are impressed with the methodology for reasons of *efficiency*, rather than *empowerment*.

Most immediately indicative of a successful approach is the fact that project staff have adopted the information exchange workshop as a replicable model for planning research with farmers. Community workshops – where farmers present and evaluate their experience as a prologue to others planning new trials – are now regularly organised without external facilitation.

The enthusiastic response of the CIAT project team has contributed to an institutional review of participatory methods. CIAT has experienced a plethora of 'participatory' methods introduced through various projects in the last few years, with some inevitable differences of approach, and CIAT's response has helped to internalise that experience. The project specifically contributed understanding of the need for flexibility in participatory planning, and that participation means more than merely conducting a PRA. In Spanish, even more than in English, participation has two senses. Passive participation, meaning to attend a meeting, or join in group activities, is the more common interpretation in Spanish, whereas in English it has acquired the more active and political interpretation implying control over decision-making. The current popularity of PRA methods encourages a 'blueprint' approach and an idealised view of its appropriateness. The research team found that it could not just repeat patterns for PRA used in other circumstances, and was also surprised to find that similar tools could be drawn on to give more control to farmers in planning and evaluating trials.

By involving NGOs in parts of this review, lessons have been shared, particularly those relating to participatory monitoring and evaluation. The failure to address community management issues can, however, be attributed to weak collaboration with the NGOs in the area. After the initial workshops, when the information-sharing role for the NGOs was clear, they have maintained an interest in the project and have participated in monitoring meetings, but have not been proactively involved with community action. The NGOs already have their own agendas and, while interested in improving rural livelihoods, are clearly still waiting to see the effects of this project before making drastic changes to their own programme. The project budget was small, and the CIAT project team had to choose between communicating with farmers and communicating with NGOs; consequently it has missed opportunities to strengthen links with other organisations, who themselves have been cautious about getting involved because of mutual distrust among NGOs. Nevertheless, through its philosophy of linking with other institutions, CIAT has identified those with an interest in continuing the project who, importantly, want to adopt the *method* not the technology.

Through a participatory approach to research, involving information-sharing and farmer control of experimental plans, CIAT can be seen to have achieved considerable success in encouraging SWC in the temperate valleys, with more than 30 farmers planting new trials in the third year of the project. Participatory technology development has thereby won approval among scientists from more traditional backgrounds, but crucial fundamental differences remain between the reasons for approval of the methodology among different stakeholders. While international facilitators, and many local NGOs, see participation as a route to empowerment of marginalised rural populations, this is not the factor which convinces agricultural researchers. CIAT staff are impressed by the range of ideas for soil conservation which have emerged in an area where three years ago they had no specific technologies to offer, the flexibility and low cost of the technology, and the speed with which these ideas have been incorporated into farming systems and been 'transferred' from farmer to farmer. The involvement of women illustrates this attitude. Initially the project team attached little importance to involving women in the information exchange workshops, but subsequently they have seen that their experience of the environment and farming is different and relevant to SWC. Women's knowledge relates to different components of the system, and different areas of the farm. In a largely male agricultural research institution, this argument is more convincing than equity arguments for a gender focus.

If national researchers are interpreting success within the classic technology transfer model, how relevant is a discussion of knowledge theory? The fact that farmers' motives for what is perceived as 'adoption' are often quite different from those expected by outsiders is only reluctantly acknowledged, but in many ways local researchers are acknowledging the transformation of knowledge, by expressing their concern over the definition of 'soil conservation technologies' and more recently, the links to changing livestock pasture management. They have seen that workshops entitled 'Soil Conservation' do not attract participants and have begun to raise some more fundamental doubts surrounding the outsiders' perception of the soil erosion problem in the valleys. From the analytical, academic point of view it helps to see knowledge transformations rather than information flows, but for research staff who have to get on with the job, such notions are not a fundamental part of the approach: they work out for themselves, what is successful, and build on it.

7. Conclusions

With respect to the dynamics of knowledge, and the process for creating new knowledge, the Bolivian experience contributes to three areas of discussion: the need to recognise the dynamics of rural contexts; critiques of participatory approaches; and the value of internalising experience for institutional strengthening.

Our experience highlights the differences between farmers' and outsiders' perceptions of soil-related problems, their causes and effects. While scientists focus on soil erosion, farmers are addressing the more immediate problems of soil fertility and humidity. Their knowledge, including perceptions and attitudes, is formed by their experience of change, including declining crop productivity, declining availability of land, conversion of grazing lands to crop land and a shift from sheep to cattle production. Migration both into and out of the area has affected ethnic and environmental balances, and the current state of the land is unlikely to be the result of current land management practices.

The project methodology was a simple one, based on the premise that relevant knowledge already exists in the zone, and that much can be achieved by facilitating its communication. But a participatory research approach is not a magic solution for the adoption of outside agendas. Farmers have incorporated some of the new ideas into their systems, but not primarily because they want to protect the soil. The conventional model of technology transfer, enlightened by farming systems research methodologies, would say that adoption or innovation of SWC technologies will not happen unless the technologies address the problem perceived by the farmers, which may mean that a technology is incorporated for reasons other than those intended by the development agent. The creative knowledge interface provides a more helpful way of interpreting farmers' responses, and a heuristic guide to project planning, but the concept needs to be used proactively, taking into account social and power relations in the community which affect *who* has access to the interface.

The creative knowledge interface takes practical form through information exchange workshops, farmers' research plans and participatory evaluation of trials. The pragmatic arguments for this approach, not the theoretical ones, convince state researchers, and by evaluating their own experience, changing attitudes and success in their own terms, the model has been approved and incorporated into a research approach for the temperate valleys.

Note

1. 'Participatory improvement of soil and water conservation in hillside farming systems, Bolivia' is funded by DFID (UK) throughout Resources Systems Programme (NRSP), Hillsides Productions System, project R6638. The views expressed here are those of the author.

11
Devolution of Decision-making: Lessons from Community Forest Management at the Kilum–Ijim Forest Project, Cameroon

David Thomas, Anne Gardner and John DeMarco

1. Introduction

Throughout much of the tropics, colonialism has contributed to a legacy of centralised government authority over management of natural resources. In many regions this has been entrenched by African traditions of governance. A protected area network that excludes people, established at the beginning of the twentieth century, has since been added to by successive postcolonial governments (Anderson and Grove 1987). The record of this approach to conservation and natural resources management is generally regarded as one of failure (for example, Adams and McShane 1992; IIED 1994). In many parts of Africa forests are administered by a government bureaucracy that is often out of touch with local people, local opinion and local needs, and communities adjacent to protected areas have ignored or actively rebelled against imposed regulations on the use of 'their' resources, and now have negative attitudes to conservation (Infield 1989; Parry and Campbell 1992; Heinen 1993; Mkanda and Munthali 1994; Fiallo and Jacobsen 1995; Nepal and Weber 1995). Without local support, excluding people from protected areas, or enforcement of regulations on access to or use of natural resources requires high levels of patrolling and policing, which is expensive and tends to heighten tension between local communities and forest and wildlife departments. Also, since many departments are understaffed, underskilled and underfunded, levels of protection have proved inadequate.

In recognition of past failures, approaches to conservation and natural resource management are now beginning to change. Throughout much of Africa, governments are slowly devolving responsibility to local communities. The CAMPFIRE (Communal Areas Management Programme for Indigenous Resources) programme in Zimbabwe is one such example (Child 1996: see also Sibanda, this volume). Much of this process of change is experimental, however, as communities redevelop capacity and recreate institutions for resource management, and governments learn to 'let go' and allow communities the responsibility for resource management.

Devolution of natural resource management has been strongly advocated for forests, especially in Asia, and there is a vast literature on the principles and practice of Community Forest Management (CFM) (for example, Fisher 1995; Jackson and Ingles 1998; Jeffery and Sundar 1999).

This chapter focuses on BirdLife International's work in the Cameroon Mountains, West Africa, where decentralisation of forest management through 'community forests', and community forestry management has been the focus of the Kilum–Ijim Forest Project (KIFP).[1] Following a brief introduction to the project area and project objectives, in this chapter we use the KIFP as a case study to explore some of the conditions that appear to be key to the devolution-participation process.

2. Biodiversity importance of the Cameroon mountains

Most of the mountains over 2,000 m in northwestern Cameroon (the Bamenda Highlands) were covered historically with Afromontane forest, and represent a biological community that is unique in West Africa. Due to human pressures, mostly agricultural, most of the original montane forest has been destroyed. The largest (at 20,000 ha), and possibly the only viable remnant of this forest type is found on Mount Kilum (also known as Mount Oku) and the adjoining Ijim ridge. The Kilum–Ijim forest lies within the Cameroon Highlands Endemic Bird Area (EBA).[2] It constitutes the most valuable remnant of an endangered ecosystem that is of global significance for biodiversity conservation. Studies by BirdLife International and others (for example White 1981; Thomas 1986, 1987; Collar and Stuart 1988; Stuart *et al.* 1990; Hutter and Fulling 1994; Wild 1994; Stattersfield *et al.* 1998) have clearly demonstrated the importance of this montane forest as a centre of endemism. The *Podocarpus*/bamboo forest on the mountain is unique to West Africa and is a refuge for several rare and threatened species including Preuss' monkey. A toad species (*Xenopus amieti*) is endemic to Lake Oku (a crater lake within the forest) and several endemic species of plant have been identified including the orchid *Disperis nitida*. In 1997 alone, at least four plants new to science were found in the forest. Fifteen montane bird species endemic to Cameroon are found in the Kilum–Ijim forest. Two of these, Bannerman's turaco, *Tauraco bannermanni* and the banded Wattle-eye, *Platysteira laticincta*, are restricted to the forest and are globally threatened. The Kilum–Ijim forest almost certainly represents the only genuine possibility for the conservation of viable populations of these species.

3. Society and economy of the Kilum–Ijim area

The area surrounding the Kilum–Ijim forest is one of the most densely settled parts of Cameroon: an estimated 200,000 people live close enough to the forest to walk there and back in a day. This high density is attracted by the rich volcanic soils, and the relatively 'temperate' climate that allows cultivation of crops (mainly rainfed) ranging from coffee, beans and maize to Irish potatoes, onions and tomatoes. The potatoes are exported to other parts of Cameroon, as well as to neighbouring countries, and the coffee is an important cash crop, although production declined dramatically following the crash in coffee prices in the mid-1980s. Most farmers farm according to a mixed cropping or an intercropping system, with maize being the dominant crop. Since the construction of a new road in the early 1990s, the southwest side of the forest is relatively accessible. The northeast side by comparison is served only by unsurfaced roads that quickly degenerate to a river of mud during the rainy season. The North West Province is a stronghold of the opposition in Cameroon and is Anglophone, in a predominantly Francophone country.

The societies around the Kilum–Ijim forest, like those elsewhere in the North West Province, still operate on the basis of a traditional centralised political system. A hereditary ruler (the Fon) and his regulatory council (the Kwifon) head these traditional authorities. Three chiefdoms (*Fondoms*) cover the Kilum–Ijim forest, the *Fondoms* of Kom, Nso and Oku. In each case the *Fon* and *Kwifon* provide a respected leadership that is accepted by all, including immigrants. The *Fon* is the traditional custodian of the land, and in his name the *Kwifon* control exploitation of its natural resources. The main immigrants to the area are Fulani pastoralists. Some are now settled within grassland 'enclaves' within the forest, and have been living there in permanent houses for at least the last 60 years.

4. Background to the Kilum–Ijim Forest Project

BirdLife International launched a biodiversity conservation project at Kilum in 1987 and has collaborated with the government of Cameroon, currently through the Ministry of Environment and Forestry (MINEF), in implementation and funding of on-going conservation and development initiatives at the site for the last ten years.

Previous attempts made by the government to demarcate the forest had failed: the boundaries were not being respected and the forest continued to be encroached upon for farmland. Dialogue with the communities surrounding the Kilum–Ijim forest has shown that people value the forest as a source of water, wood, medicine, honey, meat and a variety of other products. But despite this, there is still strong pressure to use the forest for farming and grazing, as individuals attempt to meet their essential needs. Since work by BirdLife International in the region began, growing evidence has emerged of a community consensus that it is possible and desirable to preserve the remaining forest for future generations, while pursuing human development needs in ways compatible with maintaining the forest.

In the current five-year phase of the project, the major challenge is to put in place a durable system to protect and manage the forest in the future. Given the sheer number of people making use of forest resources and functions, the strong attachment of the local communities to the forest and the growing recognition of the role of communities in resource management, a community-based approach is considered the only viable one. The project's principal strategy is to bring about the creation of community forests, as foreseen in Cameroon's new forestry law, through negotiation between the communities and MINEF. The success of this strategy depends on building capacity in both MINEF and the local communities to develop participatory management plans, develop forest management institutions, set up appropriate management mechanisms, and to monitor forest management on an on-going basis.

Without prejudicing the negotiation process which must take place between the communities, traditional authorities and government, a flexible model of community forest management is being proposed, largely based on successful experiences in Asia, and taking account of community development experience in Cameroon. The model involves empowering and training organised groups of intensive forest users (who have the strongest interest in maintaining certain values from the forest), together with individuals from other sectors of society (representatives from village authorities, farmer groups, women's groups, and so on) to be responsible for the day-to-day management and protection of sections of the forest. The user groups will represent the community, as defined in the legislation. They will be given this role subject to an agreed management plan negotiated with and monitored by the Ministry of Environment and Forestry and the Traditional Authorities. The policy and legal framework for CFM in Cameroon is discussed in more detail in section 6.1.

5. Objectives of the Kilum–Ijim Forest Project (KIFP)

The goal and objectives of the KIFP acknowledge the links between environment and socio-economic development. They also recognise that participatory resource management doesn't occur in a vacuum, but requires strong, stable, respected and democratic institutions.

Overall goal:

- To conserve the biodiversity, extent and ecological processes of the Kilum–Ijim forest and to ensure the sustainable use of natural resources by local communities.

Core objectives:

- an effective participatory, community-based forest management system for conservation and sustainable use of the forest is in place and functioning;
- communities, traditional authorities and Government have the capacity to implement community forest management;
- local livelihoods are improved in ways which contribute to the conservation of the forest;
- a permanent system is in place for monitoring the effectiveness of forest management.

6. Multiple goals: multiple conditions: multiple activities

The approach to conservation and sustainable resource management in the Kilum–Ijim forest has evolved in the ten years in which BirdLife has been active in the area. A process approach has responded to experience, to lessons learnt and to changing social, economic, legal and institutional circumstances.

The project's goal encompasses two objectives: biodiversity conservation and sustainable development. These two objectives are complementary and achieving success in one is dependent on success with the other. Just as there are multiple goals, no single factor can be identified as instrumental to project success. Instead a range of conditions, and activities designed to create those conditions, contribute to the process. The principal conditions are described below.

6.1   Enabling policy

Communities living around the Kilum–Ijim forest have long used the forest and its resources, and use has in the past been regulated by traditional authorities. Centralised government control of forestry has eroded those rights and responsibilities. Therefore, although people have continued to use the forest there has been no formal *de jure* mechanism at a village level for regulating use and access, and Forest Department capacity (staff, skills and resources) has been inadequate for effective regulation, enforcement or monitoring. When the project began in 1987 it recognised the importance of community participation and involved local communities in setting boundaries to 'their' forest to prevent further agricultural encroachment. However, this process of community involvement had no basis in law and as a result was inherently unstable.

Legislation in itself is not enough to ensure the success of community-based approaches to NRM. In countries where the policy framework remains weak, where resources for enforcement or administration are inadequate, where there are high levels of corruption, or where commitment from government staff is lacking, legislation may be worthless. Conversely, where CFM has support from communities and local forestry officials, devolving responsibility for forest management may be possible in the absence of a legal framework. Nevertheless, legal recognition of community forestry does provide an important opening for decentralised management and provides some legal security for those communities that undertake it.

In 1995 the government of Cameroon adopted a new forestry policy. The main legal instrument for implementation of the new policy is the forestry law of 1994.[3] The policy highlights the government strategies for making the forestry sector contribute to the socio-economic development of Cameroon through engaging NGOs, economic operators and local populations. The objectives of the policy include to 'Increase the participation of local populations in forest conservation and management in order to contribute to raising their living standards.'

It states that:

> By associating the rural population in its [the new forestry policy]
> implementation, especially through the development of village
> *community forests*, the policy seeks to secure substantial benefits for
> village communities as well as motivate them to better protect forest
> cover. (in Ministry of the Environment and Forestry 1997, emphasis
> added)

Community forests are a new category within the classification of the
national forest estate, within which 'Forest products of all kinds … shall
belong solely to the village communities concerned' (section 37, paragraph 5). Basic requirements for attribution of a community forest
include a management plan approved by MINEF and an appropriate
legal entity. The management plan shall specify 'the beneficiaries, the
boundaries of the forest allocated to them, and the special instructions
on the management of areas of woodland and/or wildlife' (section 38,
paragraph 1). The community concerned must have legal entity in
the form of either an Association, a Cooperative, a Common Initiative
Group or an Economic Interest Group. MINEF is mandated to give assistance in this process, but capacity within the organisation (experienced
staff) has been a limiting factor. The commitment to local participation
in forest management that is enshrined within this law has provided an
important policy and legal framework for returning responsibility for
management of the Kilum–Ijim forest to local communities.

6.2 Capacity-building

Devolving responsibility for forest management to local communities
requires new skills and understanding at many levels. At government
level, 'letting go' of forest resources, relinquishing certain powers and
taking a role as monitors and facilitators of a process that reduces their
direct authority (and revenue-generating capacity) requires a radical
reorientation of the Forest Department's role. Implementing the law is
likely to be complex and revision of the decree pursuant to the forestry
law, specifying the conditions for implementing community forest
mechanisms, is key to this. The Department for International
Development-funded Community Forestry Management Unit based at
the offices of MINEF in Yaoundé is playing a key role in developing
capacity at national level.

At a local level, capacity must be built among the communities surrounding the forest (including, for example, women, youth, elders and

different ethnic groups); the traditional authorities; MINEF staff posted to the area; and administrative authorities. Although traditional authorities have in the past had responsibility for managing forest resources, many of these structures have broken down. Also, the boundaries to new institutional frameworks and structures for governance – as a result of processes of democratisation – do not always coincide, and yet capacity for cooperation and coordination is needed within all these institutions. The project is therefore working with traditional institutions, which are still extant, influential and well respected, within a tripartite framework for forest management (Figure 11.1). Capacity has to be built simultaneously within each of these participant groups if the process is to succeed.

6.3   Local participation

In circumstances where the forest is widely used and population pressure and demand for land is high, conservation measures must work in partnership with local people. In 1987 when BirdLife started conservation

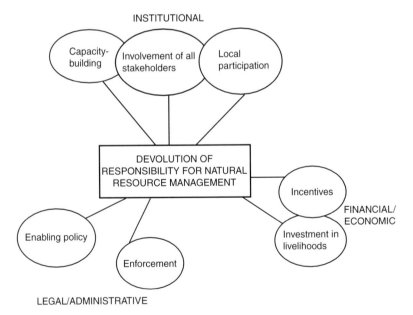

Figure 11.1 Summary of conditions for devolving rights for natural resource management to local communities

work in the area, forest destruction was severe and posed an immediate threat to the viability of the forest. Through consultation with representatives from traditional authorities and government, project staff took steps to agree and demarcate a boundary to the remaining forest beyond which no further forest would be cleared. The boundary was agreed in 1991 and since then the boundary has been largely respected, and any new encroachment has been dealt with promptly by the traditional authorities.

Whilst agreement on the boundary was a major achievement for the project, it was also recognised that more participation would be necessary if the forest was to be managed and conserved in the long term for the benefit of people and biodiversity. The 1994 forest legislation provided a mechanism for this. The process at the Kilum–Ijim forest is based on a five-stage participatory process involving:

- *information* in which the idea and processes of CFM are discussed with and within communities, allowing them to reach a decision as to whether or not to proceed;
- *investigation* of the existence and availability of resources, as well as use patterns and rules of access and use rights;
- *negotiation* between forest users and forest department officers of a forest management plan;
- *implementation* of the plan; and
- *review* of the impacts with subsequent adjustments if required.

The investigation and negotiation stages take time to identify all users of the forest and to involve them in decisions about its use, and to build rapport between forest users and the forest department. Through this process there has been a reawakening of community responsibility for the forest. Communities are taking the initiative in controlling outbreaks of fire in the forest, and in preventing the illegal extraction of forest products such as *Prunus africana* bark by outsiders (*Prunus* can only be harvested under licence from MINEF, see section 6.7).

6.4 Stakeholder groups

In negotiation of forest-related issues, a fundamental requirement is to involve all stakeholders, and to facilitate the working together of people with diverse and often conflicting interests. Agreements on rules governing access to and levels of use of forest resources may be worthless if certain stakeholders are excluded from the debate and so have no ownership of decisions. The CFM process at the project involves an investigation phase that aims to identify all the stakeholders and draw them into the process.

Graziers at the forest are a case in point that demonstrates the importance of full participation of all stakeholders. Recently, cattle and goats have been herded into the forest in contravention of a prefectorial order and against the wishes of most of the community. Attempts to resolve the issue through negotiation failed since graziers chose not to participate in community meetings. The intervention of the *Kwifon* as a neutral and respected third party enabled all stakeholders to be drawn into the meeting. Since the involvement of the *Kwifon* many goats have been removed from the forest (Gardner *et al.* 1997).

In another case the project was working through the traditional village chief (and other community representatives), but there was no progress and little involvement of the community despite expressions of interest. Inquiries then found that the community did not respect the village chief and the project had to work through the *de facto* leader instead.

6.5 Incentives

People living adjacent to the forest are poor, and need cash incomes and employment. Agricultural and grazing land is scarce. Any attempt to conserve the resources of the Kilum–Ijim forest must meet priorities of local people. Participation and devolution are processes that help to ensure that local people have a say in how the forest is managed. However, people will not support forest conservation unless they see benefits to them and their families that outweigh benefits from conversion of forest to farmland.

The Kilum–Ijim forest provides many direct economic benefits for local people. These include forest products that are sold, bartered or used for subsistence. Fuelwood, medicines, fibres, fruits and nuts, some game (mainly rodents) and timber for carving are examples. Honey, a valued product from the forest, is of high quality, and beekeepers are amongst the most active supporters of forest management. The project supports honey production through training, introduction of intermediate beekeeping technology and marketing. Because of the project's activities at least 5,000 additional hives are now in production in and around the forest. The forest also has important functions that indirectly bring economic benefit to the region. Foremost amongst these is its role in regulation of water flows. Local people are well aware of the role of the montane forest in regulating stream flow. In neighbouring areas where the forests have been removed streams are now seasonal;

those flowing from the Kilum–Ijim forest are perennial. These benefits, which are all directly related to forest quality, are important incentives for forest conservation. In addition, important non-economic factors create a desire for forest conservation. A complex web of culture, spiritual value, pride and a desire to emulate 'outsiders' creates an important force for conservation.

Benefits from some forest resources and of most functions are fairly universal. Thus most households collect fuelwood from the forest, and all will depend on water (from streams that originate on the forested slopes of the mountain). The cultural importance of the forest is deeply appreciated by individuals from all the major ethnic groups residing around the mountain (Kom, Nso and Oku). For other resources distribution of benefits is more restricted. Thus some resources are used mainly by certain economic sectors within forest adjacent communities (e.g. favoured timber species used for carving, such as *Polyscias fulva* used by mainly male woodcarvers), others by particular economic and/or ethnic groups (e.g. graziers are predominantly, although not exclusively, either Fulani pastoralists or relatively wealthy sectors of local communities) and for others there are gender divides (hunting, now restricted mainly to trapping of rodents, is exclusively male). However, some of these boundaries are being broken down by the project. For example, beekeeping, earlier a male preserve, is now practised by a growing number of women. Attention to the diversity of users of forest resources is important in developing as wide a constituency as possible in support of forest conservation.

6.6   Livelihood investments

Whilst it is important that local people value the forest, most such benefit streams are fairly small (albeit long-term) at an individual level. If the forest is to be conserved, local people must be able to make a minimum standard of living. At Kilum–Ijim this means working on land outside the forest, particularly with respect to agriculture: this is the main livelihoods activity in the area and also the biggest threat to the forest (through expansion of farmland). Many forest-adjacent people are poor and have immediate and urgent needs that they struggle to meet from existing agricultural lands or other livelihood activities. Participatory needs assessments have been carried out by the project's technicians. The project's livelihoods programme includes improved farming methods (contour bunding, live fencing), introduction of new crop varieties (for example onions, oil palms, potatoes), afforestation (fruit tree production, seedling production and nursery management) and livestock health and husbandry. By investing in these areas the project aims indirectly to reduce pressure for forest clearance, allowing the continuous and sustainable stream of benefits from the forest to be realised. It also aims to influence people's willingness and desire to keep the forest (Abbot *et al.* 1999).

6.7   Enforcement

Community Forestry aims to hand over responsibility for forest management to local people. However, rules will still be needed to control access and levels of exploitation, and for some mechanism of enforcement: 'community forestry' does not equate with 'open access'. Not only must there be a mechanism for controlling use by community members, but rules must also exclude outsiders. This is particularly pertinent in the case of *Prunus africana* bark. The bark of *Prunus* is exploited commercially for extraction of a product used in the treatment of prostatitis and its high value has encouraged an external network of outside exploiters who illegally remove *prunus* bark in large quantities.

Existing rules for forest use forbid stealing of honey, hives or trapped meat, for example, and the number and location of trap lines is regulated for much of the forest (Shepherd and Brown 1997). New regulations will also be introduced, in accordance with the management plans agreed for each community forest. These are likely to include controls on the species of tree that can be felled (for example *Polyscius fulva*, a favoured carving wood). Some forest management institutions have elected to issue members with identification cards which will legitimate their use of certain forest resources and give them some powers to report non-legitimate users.

Whilst traditional authorities have some authority to fine and confiscate illegally collected products, the role of the Ministry of Environment and Forests remains important, and one common concern of communities when CFM is proposed is that the government administration be behind them to support enforcement of their community rules. In addition, and especially with products of high commercial value such as *Prunus africana* bark, where organised networks of illegal harvesters are involved, policing and enforcement by the government will continue to be an essential adjunct to local measures.

7.   Project impact

The project has put in place a comprehensive, tiered monitoring system for the forest ecosystem, comprising satellite image analysis, fixed-point photography, permanent vegetation quadrats, point-counts (for birds) and measures of human use (numbers of traps and bee hives encountered, for example). The results from a comparison of 1987 and 1995 SPOT images are encouraging. They show that forest encroachment virtually ceased when the project began, and that since that time areas of degraded forest have started to regenerate. In one case an outlying forest fragment (a sacred forest) has become contiguous with the main forest (Maisels 1999). For the livelihoods programme, an uptake survey carried out in November 1998 found that adoption of technologies had exceeded targets set by the project. Also in 1998, an evaluation of attitudes to forest protection (carried out in association with the International Institute for Environment and Development) found that attitudes of all communities have become more positive through time, and that this was directly linked to the livelihoods programme (Abbot *et al.* 1999). The project has also had significant success in terms of institutional development. Several members of staff within MINEF now have skills and experience in participatory approaches to development and forest management. At the community level several forest user groups have been legally registered and 'common-use rules' have been drawn up that cut across the forest (that is agreed and shared by communities within the *Fondoms* of Oku, Nso and Kom). Given many of the tensions that have existed between adjacent communities this is a major success. Therefore, based on ecological, economic and social indicators, the project has had a positive impact.

8. Conclusion

Analysis of the reasons for failure of centralised, exclusionary and protectionist approaches to conservation and NRM suggest that devolution of responsibility will provide an alternative approach more suited to the demography, social systems and economy of many developing nations. The current spotlight on democratic processes and the focus of many international (bilateral and multilateral) agencies on poverty reduction (for example, UK Government 1997), adds to the appeal of approaches that empower local people through decision-making about resource use.

This chapter has reflected on the conditions that have been found necessary for devolved management of natural resources in one small geographic area, and generalisations from such a case study must be made with caution. With that caveat in mind, experience at Kilum–Ijim suggests that devolution of responsibility requires a multiplicity of conditions, and this will be reflected in project goals, and the activities required to bring about those conditions where they are not pre-existing. This is likely to put high demands on agencies involved in projects where devolution of NRM is the objective: they must tackle local institutional capacity-building, government capacity-building, local economic development, legislation, enforcement mechanisms, management planning and monitoring mechanisms. They must understand traditional rights and obligations, and make sure that all stakeholders are involved in decisions about resource use. Failure to deliver on any one of these requirements could lead to failure of the overall objective. Organisations need to recruit staff with appropriate skills and experience. Having these staff on the organisation's payroll may avoid the complexity of projects managed jointly with other organisations, but probably only the largest NGOs can enjoy this luxury. Improved networking and sharing experience between organisations with different and complementary skills may also facilitate the need for diverse skills. Willingness to share lessons and experience and to listen and learn from others (for example, conservation organisations learning from those that have more experience in development) is also important.

This case study also suggests that changing laws and building institutions is likely to be a lengthy process, taking upwards of 10–15 years. Building a relationship of trust between communities and governments has at first to break down the barriers to collaboration that have built up over years of conflict: this will not be achieved overnight. Organisations must make sure they have the commitment to see the process through to completion. The long process also needs commitment of resources over a suitable period of time. Donors who claim to be committed to participatory, devolved NRM need to demonstrate this by making funds available over the time period that the process requires.

Perhaps surprisingly for an organisation focused on biodiversity conservation, the Kilum–Ijim Forest Project has not over-emphasised the biodiversity aspect: indeed, it could be argued that, until recently, biodiversity survey and monitoring has not received the attention it deserved. Instead, the *diversity* of project activities has been the biggest problem. Staff have tackled areas with inadequate experience, and without suitable networks, or communications to networks. As a result, activities have either been neglected or (now familiar) mistakes have been made. Although the overall project objective has remained largely unchanged, project activities have tended to reflect the skills, interests and strengths of the project manager. A change in the project manager has therefore led to a change in project focus, with consequent confusion amongst communities and local project staff. More recently, with an improved planning process, the project has been more consistent in its approach, regardless of staff changes. This has partly been achieved by the appointment of specialists, on medium- or short-term contracts, to fill skills gaps, as well as better communications (for example e-mail links).

Although the project has maintained a programme continuously since 1987, funding has been a recurrent difficulty. The periods at the junction of phases (every three years) have been very disruptive as funding shortfalls have required the suspension of activities and laying-off of staff. Whilst some donors are now beginning to make longer-term funding available, most are not, requiring NGOs such as BirdLife to invest resources continuously in fundraising.

This case study shows that whilst participation and decentralisation of decision-making is often regarded as the panacea to many of Africa's environmental problems, it requires very significant capacity in the organisations that undertake it.

Notes

1. BirdLife International is a global partnership of conservation organisations that works in over 100 countries. BirdLife International aims to conserve all bird species on earth and their habitats, and through this, to work for the world's biological diversity and the sustainability of human use of natural resources. BirdLife International implements the Kilum–Ijim Forest Project with the Ministry of Environment and Forests. It receives support from the Department for International Development (Joint Funding Scheme), the Global Environment Facility–World Bank, world Wide Fund for Nature (Netherlands) and the Netherlands Government (Ministry of Foreign Affairs & Ministry of Agriculture, Nature Management and Fisheries) in the framework of their Programme International Nature Management (PIN) 1996–2000.
2. Endemic Bird Area is the term used to describe areas with two or more bird species with restricted ranges (total range size $50,000 \, km^2$ or less) (International Council for Bird Preservation 1992; Stattersfield *et al.* 1998).
3. Law No. 94/01 of 20 January 1994.

12
Changing Natural Resource Research and Development Capability: Whither Social Capital?

Stephen Biggs, Harriet Matsaert and Adrienne Martin

1. Introduction

This chapter explores changes taking place in natural resource projects and the usefulness of the concept of social capital in understanding these changes. Our definition of social capital is of 'networks norms and trust that facilitate coordination and cooperation' (Harriss and Renzio 1997: 922).

In discussing the conditions for success of agricultural extension pro-grammes, emphasis has often been placed on identifying networks, linkages, systems of cooperation and trust amongst the client group of a project (often defined as poorer farmers). Less often stressed is the need to understand the linkages in the wider environment: within and between institutions, the commercial sector, NGOs, etc. and between all these and the client group as a prerequisite to achieving research and development capability for natural resource management. For the purposes of this chapter we define research and development capability as: 'the capacity to provide rural people (particularly poorer, disadvan-taged groups) with sustainable livelihood options.' We distinguish here between research and development capability and institutional capac-ity. The former refers to the research and development processes under-taken by formal and informal institutions. Institutional capacity refers to the formally established institutional organisations (agricultural research institutes and so on, and formally constituted NGOs) within which that research and development capability operates.

We consider four case studies, drawn from the authors' field experi-ences, which we believe illustrate 'successes' in building research and development capability. In the tradition of participant observation research, we reflect on the lessons that can be drawn from situations in which we have worked, and we describe examples where individuals within institutions sought out and found existing linkages or shared interests with which they could work. Just as they experimented with participatory methods for working with farmers, they also experimented with methods of supporting learning and action in arenas beyond the village.

In the final section, we explore some common themes that can be drawn from these experiences. Do they tell us how research and development capability can be built? Do our existing methods of project planning, monitoring and evaluation place sufficient emphasis on the strengthening of research and development capability as opposed to short-term products and the building of formal organisations and structures? Finally, is the concept of social capital a useful one in assisting us to describe and evaluate the linkages and coalitions, which underlie this research and development capability?

2. Case study 1 – Changing research capability to meet the needs of communal farmers in Namibia

When Namibia attained independence in 1990, its agricultural research system was forced to change dramatically. Prior to independence, research had been mainly targeted towards a minority of commercial farmers. Post-independence policy stated that research should now be targeted to the majority of farmers: those in areas of communal land tenure with mixed farming systems primarily for subsistence rather than commercial production. New approaches to agricultural research were piloted in different parts of the country. The Kavango Farming Systems Research and Extension (KFSR/E) team, formed in October 1994 to pilot a farming systems approach to research and extension in Kavango region, was situated in the Directorate of Research. The gap between research and extension was bridged by drawing team members in equal numbers from the separate Directorates of Research and of Extension, both in the Ministry of Agriculture. The regional project coordinator was the Chief of Extension for the region. Significantly, all staff members remained in their respective line jobs. Therefore, the project was firmly embedded in the ongoing institutional structures in the country. The team received funding and technical assistance (one agronomist and one social anthropologist technical cooperation officer) from the UK Department for International Development (DfID) for the first three-year phase.

KFSR/E's purpose was to develop regional capability for adaptive research, extension and training through a farming systems approach.

It was committed to participating with farmers, focusing on the needs of the poorest farmers in the region. Much of KFSR/E's work involved identifying existing organisations and channels of communication ('social capital') and finding effective ways of working with rural dwellers in Kavango. Through ongoing studies the team strove to identify and work with existing organisational structures and social networks both for technology development and the dissemination of information. Where this was not possible, new structures and forms of 'social capital' were developed. In some villages farmer groups were formed specifically to carry out agricultural research with the Ministry of Agriculture (represented here by KFSR/E).

An equally important part of the team's work was in creating links with the relevant government institutions and NGOs, to allow rurally-based groups to influence research and extension processes and policy. As a result of the team's efforts, and because the project was initiated at a favourable moment, KFSR/E had considerable success in building these links with extension and research. At the end of the project's first phase, research and development capability had been built in several ways:

- Attitudes of senior researchers and extension staff (and funders) had begun to change from seeing 'project success' in terms of seeds distributed, yield increases in fields, and so on, in the short term, to one of effects of the project on sustainable local research and development capability and changing national research and extension agendas.
- The Directorate of Research, which was centrally managed, was beginning to move towards a more regionalised approach. Directors of plant and livestock research sections had agreed to base their national planning on needs identified by communal farmers (identified by farming systems research teams). Some commodity researchers were beginning to take some of their experiments to the village and were allowing these to be evaluated by farmers.
- The Directorate of Extension was committed to develop a farming systems approach. For them this meant moving away from on-farm 'demonstrations' to a process of participatory systems analysis, needs assessment and learning through on-farm experimentation.
- Attitudes towards farmers in the northern communal areas were beginning to change. Research and extension personnel were beginning to see farmers as active experimenters rather than passive recipients of information, and to form partnerships with them. The overall result was the creation of an environment in which formal institutions, farmers and informal groups were beginning to work

together, resulting in an improved research and development capability in Kavango.

How did these changes come about? Much can be attributed to the national political situation and a shared vision that things must change in the new Namibia. Participatory research and extension approaches fit well with the post-independence emphasis on democratisation. The project had support from top-level managers in research and extension. The main resistance to change was found amongst research and extension scientific and field staff.

The team used various strategies to encourage government research and extension staff to explore the advantages that farming systems and participatory approaches might have for them and to change their working methods to incorporate these new ideas. These included detailed institutional analysis, used to identify existing strengths or interests in the Ministry which KFSR/E could work with and to pinpoint key decision-makers and the use of 'events' as a forum for learning and action. Each is discussed in more detail below.

2.1 Institutional analysis

At an early stage of the project life (March 1995), team members carried out a regional survey in which they identified all local organisations (both government and non government) whose work had some relevance to the project objectives. The study also helped the team become aware of existing 'social capital' (networks, norms and trust) within and between organisations that might help or hinder the project's activities. Later in the year, representatives from many of these organisations were invited to a project advisory meeting. A participatory project planning approach was used to carry out stakeholder analysis and to redefine the project's goal, purpose and outputs. From an early stage the team attempted to involve all those who had a potential interest in the project's work, and to encourage them to feel part of it. This initial analysis was important at a later stage to help the team identify who should be invited to key 'events'.

2.2 Identification of social capital that can contribute to building research and development capability

When the KFSR/E project was initiated, most Namibian research work was being carried out without reference to the main target group – the northern communal farmers (ISNAR 1993). In the Directorate of Research, strong networks, norms and trust existed amongst researchers within the well-equipped and financed research stations, and between researchers and commercial farmers. The social capital that had developed around the pre-independence apartheid system was, however, the wrong type for current development aspirations! Because black Namibians had limited access to education in Namibia, most researchers came from the elite white sector and had strong links with each other and with the community from which they came (the commercial farmers). After independence, these networks remained strong (the researchers on the whole were not pro-independence and have not necessarily accepted the new ideologies). The links required to create strong research and development capability in the northern communal areas, amongst previously disadvantaged farmers, did not exist. Researchers had few contacts with their extension colleagues in the northern communal areas, because these extension personnel came mainly from the communal farming population. Links between communal farmers and researchers were pretty much non-existent.[1] The legacy of inequality, exploitation and mistrust between ethnic groups made the formation of links between these potential 'research partners' very problematic.

An understanding of the institutional history of the national research system and its social capital helped the team to go beyond the very negative scenario and to identify positive elements: linkages, motivations and experiences that could be used to build new types of social capital for a changing research and development capability. One example was in the team's interaction with the Horticultural Research Unit in the Ministry of Research. After independence this unit was targeted mainly on the Kavango region where horticulture was seen as one of the most promising enterprises for food security and income generation. There was also a small region of commercial horticultural production (30 German immigrant farmers) around a town 300 km south of Kavango. The main horticultural researcher based at the horticultural research unit shared a similar background to the German immigrant farmers, and had strong associations with these commercial farmers.

In common with others in the research system, the upbringing and experience of this researcher made the prospect of working with communal farmers in Kavango unfamiliar and daunting. He felt that it was easier for foreign researchers without his social background to work with black farmers. The researcher was required by the Ministry, however, to work in the communal areas. In 1995 a number of trials were set up at the Kavango research station, and their management delegated to a local research technician. The researcher visited the trials every month or so. There were few links with farmers, though occasionally extension staff were invited to bring farmers to the station to view the trials. This situation was very unsatisfactory: the experiments were not using the varieties of vegetables that farmers themselves were growing. Moreover, the varieties tested in the research station were not available in local shops, which frustrated farmers who visited the trials because of their interest in a particular variety. At the 1996 regional research planning meeting the team urged the researcher to consult and if possible involve communal farmers in the research work. This prompted a defensive response, that farmers could not be trusted to manage trials properly and that he had no time to carry out interviews with farmers in the region.

Despite the reluctance of the horticulturist and most other commodity researchers to participate in the farming systems approach, project team members made efforts to improve their understanding of the research directorate, and to build linkages by attending the monthly plant production research meetings. In 1997 the project's team leader found a way forward. He had observed the close relationship between the horticultural researcher and the small community of commercial horticultural producers. He suggested to the researcher that they might jointly undertake a participatory survey with these producers. This was a great success, resulting in the identification of several participatory technology development areas. In carrying out the survey, the horticultural researcher gained confidence in the use of participatory methods. Also the team leader gained the trust and friendship of the horticulturist. In collaboration with the team leader (and KFSR/E), he decided on some on-farm work in the communal areas. In the 1997–8 season several technologies were offered to communal farmers for on-farm testing. The horticulturist also initiated research activities examining the role of root crops in the Kavango farming system.

By helping to create these new linkages between communal farmers and the horticultural researcher the Kavango team felt that they had strengthened the local (and national) research and development capability. If the Kavango project was wound up tomorrow, these linkages would probably continue. More importantly, the changes in perceptions of the horticultural researcher and the farmers collaborating in the work would endure.

2.3 Identification of key actors and decision-makers

As the team deepened its understanding of the research and extension directorates, team members were able to identify key individuals and decision-makers in a position to bring about change. These individuals were invited for training seminars, project visits and to participate in research activities. The clear support of key decision-makers enabled field staff to start using farming systems and participatory methods, even where some of their colleagues were quite negative. The supportive political and policy environment was a critical factor in changing research and development capability to meet the needs of communal farmers.[2]

2.4   Events as a forum for learning and action

Several events were critical in gaining institutional support and developing linkages for participatory farming systems research and extension. One was an annual regional research and extension planning meeting initiated in July 1996. This meeting was the first forum in the history of Kavango where researchers gathered to listen and respond to the requests of farmers and extension personnel. In 1996, the first year that this took place, the meeting highlighted some important requests from farmers and extension, but they were frustrated that researchers did not respond to them in the next season. The meeting came too late to affect the next season's planning in all but a few cases, but the effects of the meeting were seen clearly the following year. For the first time, some researchers prepared their research proposals to meet the needs which had been expressed by farmers and, perhaps more importantly, aware that they would have to present and defend their proposals in front of farmers. Through this forum (now an annual event) the behaviour of researchers began to change without any need for others to criticise their methods or research activities directly. At the public forum researchers were able to learn and to change their actions without losing face as individuals.

A second important event, the 'Kavango Seed Fair', was held in November 1997 in response to the continuous complaints by farmers that many types of indigenous seeds had been lost (the last few years had been very poor rainfall years). The Fair followed a good rainfall year when crop yields had been high. The KFSR/E team knew that seeds had been lost in some villages and not in others and hoped that this event would allow some redistribution of seeds to the areas where they had been lost. Farmers' groups from every part of the region attended the meeting, bringing samples of all the seeds harvested in their village that year. Groups were encouraged to exchange seed, and prizes were given to those that had managed to retain the greatest biodiversity in their villages. As well as inviting farmers, KFSR/E and extension personnel (the hosts) invited many other institutions with an interest in or a contribution to make to seed supply. Researchers brought samples of the seeds that were in final tests or were ready to release to farmers.

The government-appointed seed supplier came with samples of next season's seed. The national herbarium brought samples of indigenous germplasm that had been collected in past years by the International Crops Research Institute for the Semi Arid Tropics (ICRISAT).

The event resulted in learning and action far beyond the team's initial aims. Researchers and extension personnel who attended were astounded by the enormous number of seed varieties that villages still maintained for each crop. This clearly showed that farmers were far from being the passive, unskilled people they had often been portrayed to be by many research and extension personnel. Ample evidence was available to refute the commonly made statement that 'these people are not capable of carrying out on-farm research'. The event changed attitudes in a way that three years of the project team telling researchers that farmers were actively experimenting and that indigenous knowledge was valuable had not achieved – yet nobody had to lose face or admit they had been wrong. In her speech at the end of the meeting, the head of the National Herbarium told participants, 'this is the most exciting day of my life so far'.[3]

The event initiated some important action. A leading millet breeder, formerly uninterested in farmer participatory research, collected samples of all 50 or so indigenous millets and expressed an intention to plant these at his research station and to invite Kavango farmers to come and discuss the different characteristics of the crop with him later in the year. Through a key area of interest, links between the researcher and farmers were formed, creating a research and development capability which may be used for other research initiatives in the future.

The members of the farmers' groups from across the region noted which groups were growing what seed. Extension technicians later received many requests to make linkages for seed exchange, which could be an important beginning to linkages and joint action and empowerment of groups who had so far been working in isolation from each other. Attempts by KFSRE to link these groups formally had been vetoed by extension in the past. New links and networks, perhaps generating new types of social capital that will have a life of their own, are likely to be very different from networks of the past.[4]

3. Case study 2 – Researchers take control of their own learning: maize research in India

Maize research in India has a long tradition. The All-India Coordinated Maize Improvement Project, part of a mature formal agricultural research structure that comes under the auspices of the Indian Agricultural Research Council, started in 1957. In 1978, researchers – part of the All-India maize project at the Pantnagar University in Uttar Pradesh – started their own on-farm participatory research programme (Agrawal *et al.* 1979). At the time, the idea of on-farm research was not novel, and several All India coordinated projects had on-farm trials and well-developed extension demonstration programmes. One reason for the initiation of the Pantnagar on-farm research programme was to investigate why improved maize varieties from the research station were not spreading to any extent in the 5.7 million ha of monsoon period maize in Uttar Pradesh.

The programme – organised by the much respected head maize plant breeder at the University – consisted of on-farm trials, surveys and different types of meetings and interactions of researchers with farmers. In the first year many of the on-farm trials were flooded and gave very low yields. Some members of the project felt that these 'failures' should not be written up, but the breeder and his colleagues went ahead and wrote up the result of trials, surveys and meetings. In what must have been a classic 'on-farm research' report, they analysed their findings and made recommendations to themselves and others. This report is even more significant when one reflects that a very senior India maize plant breeder was extremely critical of the researchers for being involved in such 'extension' activities and being so 'unprofessional' in publishing the results of such poor trials.

Their recommendations included suggestions for changes in extension advice, changes in the on-farm programme for the next year and changes in the priorities of the breeding programme at the research station. One suggestion concerned greater emphasis on breeding for stress tolerance. This had immediate implications, and the breeders started to strengthen measures to screen out material without stress tolerance. They found that they frequently had heated discussions with farmers in the field over what characteristics the breeders should select for. The researchers also became more careful about making claims about the relevance of their varieties for the difficult growing conditions of Uttar Pradesh. Significantly, the flood in the first year reminded them that they were essentially breeding for 'unflooded' farmers' conditions rather than for all maize growing conditions. In the first year the researchers also found that socio-economic situations of richer and poorer farmers were very different in such things as labour availability, which had important implications for the type of agronomy research that might be relevant for addressing their problems (Biggs 1983).

Many more things were either learnt (or 'relearnt', because some of the issues coming up were not always 'new', but had been forgotten or had not figured prominently in the past in guiding future research station activities). In the course of the programme, the researchers also found that they built up good working relationships with some farmers. For example, some farmers offered to store their equipment rather than the researchers having to take it back after each visit to the research station. In this programme, the traditional divide between the social scientist (doing surveys) and the natural scientist (doing trials) did not occur. The natural scientists often directed what information they needed collected in surveys, and were also involved in the gathering and analysis exercises. Significantly, the maize breeder used the results of his surveys in discussions with well-meaning visiting international maize breeders with a great deal of experience of working on maize improvement problems in India, to defend his research decisions in the face of alternative suggestions for what he should be doing.

Although the on-farm research was funded by a very small grant from the International Centre for Maize and Wheat Improvement (CIMMYT), the greatest formal research input was the time and other human resources of the senior and other researchers from Pantnagar. They invested their time in basic research activity, continually appraising the nature and implications of 'the farmer problem' they were mandated to address. Changes in the research station programme took place, because the head of the research station activities was highly involved in the on-farm programme and not only learning but also acting on the information he and his colleagues were gathering and analysing. Some of the learning was reflected in the on-farm report, but other types of knowledge were also assimilated. The 'on-farm' project enabled the Pantnagar programme to learn, not just because the head of the maize plant breeding programme wanted to learn himself, but also because he had the flexibility in his own programme to implement the implied changes in his breeding programme. This was not an easy path. Many problems arose, and the team were criticised. For example, those in the formal extension programme wanted to know why these 'researchers' were doing 'extension' where they had the extension mandate. To some extent, however, these issues were resolved when the research nature of the on-farm research programme was explained. Many issues and problems – now so well documented for participatory research activities (Okali *et al.* 1994) – emerged in the project, and were (significantly) addressed as they went along. The on-farm project continued for at least three years (Biggs 1983).

This case illustrates that while there might be formal structures (networks, linkages and so on) for the two-way flow of information between farmers, researchers and extension agencies in a mature research system, these do not always result in effective flows and changes in the activities of actors (i.e. in social capital). The formal system in this case was flexible enough to allow the initiation of the on-farm programme. The agency of the team leader and his team was particularly important in this endeavour. What started as possibly a 'one-off' event continued for several years and increased the research and extension capability in this area of maize research.

4. Case study 3 – A learning event for triticale researchers in the Himalayas

One of the exciting developments of the 1960s was the possibility that a new 'scientist-created' crop (triticale) could help poor farmers in the harsh conditions of the Himalayan hills (Srivastava 1973). Researchers had made a very difficult cross between two species (wheat and rye). The potential of the new crop derived from the hardiness of the robust rye plant and the grain quality and yielding ability of wheat. International donors paid for a triticale programme in Mexico, and augmented the triticale breeding programme at the Pantnagar University. This conducted multi-locational trials both on the plains and at research stations in the Himalayan hills. One of the main breeding objectives was to improve the grain quality of triticale. The on-station research led to farmer demonstrations in 1974–5 and a subsequent programme of extension minikits. By 1977 reports suggested that triticale was spreading fast and that science had made a significant contribution to poverty reduction in the Himalayas. In 1978 an on-farm research programme of varietal and agronomy trials was designed to help in this research/extension process (Biggs 1982). A survey of triticale users was designed to get feedback from farmers.

An international conference was organised in the hills, to take place when the triticale and other winter crops were about to be harvested. The on-farm and on-station research were prime features for the international conference. What happened at the conference was salutary for the researchers. It emerged that the current generation of triticales did *not* actually have major advantages over competing wheats and other crops. Farmers were *not* particularly concerned with the poor grain quality of triticale, as compared to wheat, as they were more concerned with high reliable yields. Some triticale varieties showed signs of sterility under certain growing conditions. Some researchers (both locally and internationally) were surprised by this and recognised that serious physiological issues needed looking into. When visiting the on-farm trials other scientists were surprised to see local wheat and barley in adjoining fields performing so well. The proposed survey of triticale farmers had been difficult because very few, if any, farmers were continuing to grow triticale after being involved in past demonstrations and minikits.

The on-farm participatory programme of trials, surveys and meetings with farmers, and the international workshop was a significant event for the researchers involved. In the past, peer group reviews of the triticale programme took place during the off-season period. Past results and future programmes were discussed in the conference rooms of research institutes, often on the plains. In contrast the international conference was set in the Himalayan hills where the triticale varieties were expected to spread. While some discussions took place in a conference room, all delegates interacted with farmers and saw triticale and other competing crops growing in farmers' fields. This not only facilitated an exchange of information between researchers and farmers but it also led to an exchange between researchers themselves (both local and international) which had not happened before. The discussions and peer group review was very different from the interactions at the normal annual off-season conference. A whole range of new information therefore became available to researchers.

To some extent, collecting and analysing the data from the on-farm trials in a formal way was unnecessary. Those participating in the conference could see the way the new triticales were performing in relation to other crops. Much 'learning' by the researchers took place in those few days. The past breeding priorities of the research programme, while technically sound, had been misdirected. The farmers' 'problem' had been misdiagnosed and no methods had been introduced into the programme to check this early research decision. After many years, the international workshop around the on-farm research programme provided a forum for 'basic' questions to be asked by research colleagues.

An event such as this international conference (together with the on-farm research) can provide a location for significant learning on the part of researchers. In this regard the on-farm research and associated conference was a major 'success', but, for some of the researchers, this led to a particularly difficult time, as the funding for the programme was greatly reduced in subsequent years. The research effort would almost certainly have been more efficient had there been a mechanism for continuous participatory interaction between poor farmers and researchers from the very start, and if researchers had had adequate information to conduct peer group reviews in an ongoing way. If such mechanisms had been in place, the researchers in the triticale programme (both locally and internationally) would have had a chance to learn, in a less painful way, how inaccurate were their assumptions about farmers' problems and possible solutions. They may well have changed the priorities and content of their research programmes at an earlier time. The triticale programme in the Himalayas had been nested in a mature national/international research system. The programme was part of the highly organised All India Coordinated Wheat Improvement Project and had strong formal linkages with the wheat and triticale programmes at CIMMYT in Mexico. This mature formal institutional structure with its system of networks, linkages, coalitions, and so on (which by some criteria might be classified as a high level of social capital) had not, however, resulted in a strong research and development capability.

5. Case study 4 – Beyond the community: developing tools and tillage systems in dryland areas of Kenya

The Dryland Applied Research and Extension Project (DAREP) worked in participatory technology development with farmers in Mbeere and Tharaka Nithi districts of Kenya from the early 1990s. This is a semi-arid area and an important part of the work involved developing tillage systems which could maximise water availability for crops in the soil. In 1993, the researchers working on soil and water management formed farmers groups at two of the DAREP experimental stations, specifically to look at these issues. Having analysed existing social capital in the area, the researchers concluded that the formation of new groups of people specifically interested in soil and water management was the best option in this situation. Though group members came from different villages and had different types of farming system, they worked well together, learnt from their differences and, in evaluation sessions, commented that this 'learning from each other' was one of the things they appreciated most about the groups. One group later adapted an existing form of cooperation, rotating savings groups, to the new purpose of allowing members to purchase tools.

As far as participatory technology development was concerned, the groups had a number of successes. Group members assessed different practices used by farmers in the area, as well as practices from other regions (observed in study tours, or introduced by the researcher) and tested those they felt were most promising on their own farms. Many farmers 'customised' the technologies to fit their own particular conditions or to improve them further and, at the end of the first season, results were assessed by the whole group. Promising technologies and new ideas were then retested and by the end of the second season the group felt ready to begin sharing their findings with other farmers in the region. Presentations by the groups became an important part of the regular open days at the DAREP experimental sites.

The farmer groups worked well for participatory technology development in soil and water management (the researchers' specialism). Problems arose when the research process led to the groups becoming interested in developing suitable tools to carry out the tillage operations, however, because in Kenya mechanisation comes under a different part of the Ministry of Agriculture from soil and water management. Earlier, tools for tillage and planting had been introduced to farmers as potential options for carrying out tillage operations. Most of these tools had been manufactured by an International Labour Organisation (ILO)-funded project, whose mandate had been to develop appropriate tools for the dryland areas of Kenya. A local artisan had designed another tool. The tools were introduced to the farmer groups for testing.

After their first season of testing tools, the farmers in the tools and tillage groups had many ideas about how to make the tools more appropriate for their conditions. Some farmers were interested in a wooden beamed plough, light enough for use by donkeys. Others wanted a planter that could be adapted for millet. Some farmers tested a hand-weeding device and had ideas on how to improve it by changing the blade. They were interested in buying some of the tools and wanted to know where they could be purchased, and what they could do about finding spare parts. This was a problem for the current researchers. The ILO consultants who had designed the tools had finished their contracts and had left the area. The Agricultural Research Institute to which the current researchers belonged had no mechanisation specialists and did not see mechanisation as part of its mandate. The rural technology units, responsible for this work, had no funding for transport, workshops or any other inputs. To involve tool manufacturers from the local town was also problematic, because the Research Institute could not allocate funds to 'non-agricultural' activities such as transporting tool manufacturers to a meeting with farmers. The necessary social capital or facilitating networks for developing tools did not exist. Without these links, despite the strong and innovative farmer groups, there was no research and development capability to develop appropriate tools (Sutherland *et al.* 1995).

While 'social capital' existed in one area of research and development (soil and water management), this was not available for use in a closely related topic (mechanisation). Even this simple example illustrates that social capital has little value as a concept of 'productive investment capability' unless the socio-economic environment is specified. In this case, the researchers were extremely lucky to find an NGO (an ILO-funded Farm Implements and Tool project) which had funding available specifically for participatory tool development activities. Thanks to this NGO, linkages between farmers and tool manufacturers were formed.

Researchers invited to several workshops (facilitated by the NGO) people who they felt could respond to the farmers' ideas. Local blacksmiths, tool manufacturers from the local town, the inventor of one of the tools, the local Rural Technology Development Unit and the engineering faculty of the University were all invited. During these workshops it became clear that the only group interested and able to get involved in creative tool development were the local manufacturers.[5] At this stage the patron NGO provided financial support for continued work with the local manufacturers, who received technical support and some materials and developed a number of tools in response to farmers' requests. Tools were exhibited some months later at a Tools Bazaar held at one of DAREP's experimental sites. The Bazaar was attended by many farmers from the area, and provided an opportunity for the local manufacturers to advertise their goods. Interestingly, some manufacturers not only responded to requests by the farmers' groups, but, as a result of observing farming conditions on previous visits, brought with them additional tools they felt would be appropriate in the area.[6] A panel of farmers consisting of men and women from the farmer research groups judged the tools. Draft animal tools were tested on the site using donkeys and oxen.

The event allowed for the raising of several important issues related to the development of new technologies for the area that could not be dealt with by DAREP, such as transporting tools from the town to the DAREP villages, obtaining credit and guaranteeing quality. Developing tools is a complex and difficult problem, particularly in a dryland area where farmers are cultivating low-value crops under very risky conditions and are unwilling and often unable to make capital investments. However the tool workshops and bazaars definitely helped by bringing people together, facilitating the formation of important links (which could continue after the project life) and allowing vital related issues such as supply and credit to be addressed.[7]

6. Discussion

Our field experiences have raised a number of important issues to be considered.

6.1 Social capital – is this a useful concept for planning and evaluating changes in research and development capability?

An important issue here is whose social capital, who creates it, who owns it, who participates in networks and who is excluded? The very term 'social capital' implies that it is like physical capital. Like physical capital, does having more social capital mean that you can produce more? As with so much of physical capital it is important to look at the specific nature and quality of the capital. In addition and more importantly, social capital has little meaning and is a misleading concept, unless one looks at the social and political context. In this chapter we have tried to relate the concept of social capital to the specific circumstances of particular research and development systems.

The case studies show that social capital is not something neutral or universally beneficial. In Kavango, the social capital involved in the close-knit network of Afrikaner researchers and commercial farmers actually undermined new development aspirations. The very different histories and ideologies of the various development actors in Kavango made the idea of strong crosscutting social networks unrealistic. Crosscutting coalitions could, however, be formed around key issues, such as the seed fair, where different actors all have something to gain. For those who seek to change research capability, identifying the issues around which coalitions can be formed may be a central preoccupation in the creation of new social capital.

All the case studies reflect 'coalitions' which went well beyond the village. Can we conclude from this that the very nature of social capital is that of coalitions which form and change depending on the shared interests of their members, and the power of the group in its own political arena?[8]

In the case of triticale, the programme was embedded in a mature institutional context where annual research conferences took place regularly, results from trials were published, and so on, with many indicators of the strong institutionalisation of science (that is, high 'social capital') in this part of the research system. However, as the on-farm programme and international workshop of 1978 showed, this investment in social capital in science produced few results for poor farmers in the Himalayas. Institutionalised 'science' to help poorer farmers had taken place for many years and at least some of the goals of scientists had been achieved. From another perspective, however, the triticale example represents a situation of a low level of social capital in terms of research and development capability because poorer farmers' problems were inadequately identified and addressed.

6.2   Building research and development capability

Much of the literature on institutional building emphasises the establishment of research sites (both in the natural science and social sciences) with buildings for laboratories, libraries, and so on. Institutional development plans concentrate on staff development programmes, on training (from the lowest level to PhDs), and on establishing formal institutional structures such as annual conferences, workshops and journals. However, research and development capability is about changing attitudes and reward structures and about changing policies. These affect the ongoing dynamics and power relationships within and between the formal and informal institutions of science. Issues such as the number of PhDs, vehicles, and so on, may be of some relevance, but how these skills, resources and networks are used is crucial.

An important topic is the issue of institutional change. Non-development literature on institutional capability often suggests that institutional sustainability should be judged by 'an institution's capacity to learn' (Brown 1998: 55). In the Kavango and Indian cases, top management allowed and facilitated change.

The experiences of these case studies highlight the importance of giving all development actors the opportunity to own their learning. One promising way to achieve this is through an 'event' where learning can take place without humiliating those who needed to change. Such events are especially valuable where power structures are very difficult to change and development actors are interested in going beyond rhetorical or short-term adoption of new methods and approaches.

In the example of maize research in India, the learning from the on-farm research could influence the on-station work without ever entering the public domain and possibly embarrassing researchers. However, even in this case the maize research team attracted sharp criticism from many of their senior peers for working in villages with farmers. In contrast to the maize case, KFSR/E's on-farm work with millet had little influence on the on-station research programme. Different people carried out on-farm and on-station work. The on-farm results, with important implications for the whole millet programme, were documented and circulated to the research community, but the lead researcher (who had not been involved in on-farm work) felt threatened and behaved defensively, refusing to adapt his programme in any way.

6.3   The importance of 'one-off events'

The importance of 'events' has been documented in the anthropology of development literature, for example the 'critical arenas' for change in state–peasant relations in Costa Rica (De Vries 1992: 53), but the nature and significance of these activities in natural resource research and development literature has been little documented. We may, perhaps, ascribe too much importance to the significance of 'one-off events', but suggest that institutional capability can only be assessed by looking at the nature of 'one-off events' as well as at the quality and nature of regular (institutionalised) events. In Kavango the seed fair could be turned into a regular 'institutionalised', event, but after a year or two it might merely become a forum for extension agents or a village fair where few scientists attend and learn. In this case the fair would have been institutionalised, but various actors would be using this formal institution for different purposes than the original 'one-off event'. Any definition of institutionalisation must include the possibility that a strong capability is characterised by the repeated occurrence of 'one-off events' that challenge existing institutions and bring about changes in the practice of science.[9]

An important criterion for evaluating projects is replicability. How will technologies and methods be more widely used? How does this relate to the concept of 'social capital' and our case study material? Our case studies suggest that 'going to scale' or repeating 'one-off' events elsewhere involves issues that need to be handled with great care. Whether the seed fair, in the spirit in which it was conducted in 1997 in Kavango, will be replicated in Kavango, or in other regions, in subsequent years will depend on the agency role of development actors and the local social and economic context. The triticale case shows how the 'replication' of formal scientific institutions (procedures, annual workshops, networks, and so on) over many years had not resulted in basic issues of the direction and efficiency of the breeding programme for poorer farmers in the Himalayan hills being addressed.

Projects, whether they are special projects to develop replicable technologies and institutions or just 'ordinary projects' are always part of an ongoing social and political context in which some types of research capability are being strengthened and others weakened. Actors (farmers, funders, researchers, development practitioners, and so on) play different roles. One needs to look at the linkages between these actors. This more overarching view of projects, we suggest, needs to be the framework for analysis, rather than a preoccupation with concerns of replicability and a narrow view of what constitutes research and development capability.[10] This approach has much in common with Roling's emphasis on the importance of effective platforms and fora so that less powerful groups have a voice in the direction of social and technical change (Roling 1996).

6.4   The agency of development actors

A recurring theme in all the case studies is the importance of individuals[11] and teams in creating events and coalitions, which facilitated a particular type of research and development behaviour rather than another. In Namibia, the KFSR/E team clearly wanted to change the orientation and practice of the Namibian research capability. The maize breeder and his team in Uttar Pradesh were interested in changing the practice of science in their arenas of work. In Kenya the researchers working with farmers actively built up linkages between farmers and a local NGO when their own institution was unable to provide support.

These actions generally challenged the social and institutional contexts in which they took place. A key feature of the behaviour of actors in these situations was an ability to see where there was room for manoeuvre and to capitalise on this. One conclusion of this study is that individual agency and team activity, as well as social capital, are critical in building research and development capability.

6.5   Indicators of research and development capability

Some of our case studies are examples of successes in building coalitions for natural resource development. But do current monitoring and evaluation systems pick up the nature and causes of these successes? What indicators in log framework and monitoring and evaluation systems really address these more fundamental issues of capability strengths or weaknesses? We may need to look for qualitative indicators, to help us to recognise growth in an institution's capacity to learn (Brown 1998: 55). These could include looking at how much time is spent on reflection and planning as opposed to action, how an organisation deals with discordant information, and whether evaluations are used as opportunities to learn. Similarly it could be useful to create qualitative indicators which measure research capability through for example identifying coalitions around key issues. As the context is so important to the development of research capability these indicators would need to be developed in an iterative way within projects, based on analysis of institutions and existing capabilities (see further on this, Biggs and Matsaert 1999).

Issues of institutional sustainability in agricultural research have received some attention in recent years. However, as Goldsmith (1993: 195) aptly suggests, 'defining and finding measurements of institutional development is controversial and tricky'. Furthermore,

> when several agencies provide only marginal input to an organisation, to ascribe credit or blame is vain …. If a process takes a long time to come to fruition, and has initial and subsequent outcomes (all of which can pertain to institutional development), its progress cannot be measured by the pace at which certain milestones are passed. A decade or more may be needed to be sure if an institutional development project has 'worked', far longer than the usual project cycle. The definitional, attribution and temporal problems probably explain why foreign aid donors as a rule do not try methodically to review and appraise their institutional ventures, not in agriculture or in other sectors. (Goldsmith 1993: 195)

Just because this area of analysis is difficult, however, further research should not be discouraged.

6.6 Importance of institutional analysis

A general lesson that could be taken from the case studies is the importance of institutional and social network analysis of research and development capability, in its broadest sense, as a starting point for thinking about and initiating natural resource research or development activities. Understanding the different histories, discourses and motivations of the various stakeholders can create a basis for positively identifying strengths on which change can be built. These factors lay behind successes in local government and the spread of agricultural technologies in Brazil (Tendler 1997), and they need to be understood by future planners.

In briefly outlining these issues, we have emphasised the importance of looking more closely at institutional capability, and the ways in which it is defined, assessed and monitored. All natural resource projects, whether designed or not to be an 'institutional development' project, are strengthening some institutional structures and modes of behaviour and weakening others. Some types of social capital are being created and others eroded. We need to understand more about these processes and be more careful in differentiating between different types of social capital as they affect research and development capability.

Notes

1. A notable exception to this generalisation appears in the 1995/96 Research Progress Report, where a researcher acknowledged the key collaborative action of local communal farmers (Fleissner 1996).
2. Note that the changed political context of Bangladesh in the early 1970s was part of the changed environment that saw the start of the Grameen Bank in Bangladesh in the early 1970s (Biggs and Smith 1998).
3. There is a home-made video of this event with KFSR/E, Pvt Bag 2096, Rundu, Namibia.
4. Other lessons from KFSR/E and the DAREP case study experiences are discussed in Sutherland *et al.* (1998).
5. The university struck up a poor relationship with the farmers' groups. They gave the group a very 'hard sell' on a number of tools they had them developed themselves and were uninterested in the thoughts and findings of the farmers. To develop any new initiative, they told us, could take a number of years as it would have to be part of someone's MSc programme. The group asked that they should not be invited again. The local inventor did provide some feedback on the tool he had made, but was uninterested in the group's other suggestions. The rural technology unit had no funding. The local blacksmiths were working with extremely basic technology and could get involved only with some basic hand tools and with spare parts.
6. A particularly enthusiastic and skilled manufacturer not only developed an improved hand planter but also produced his own low-cost sprayer and a sprinkler system!
7. This experience is discussed in more detail in Mellis *et al.* (1997).
8. Grindle (1997) has discussed the importance of trust and networks within organisations when discussing the culture of different development organisations. We are suggesting that the analysis should go beyond the organisation and look alliances and networks in broader sense. These coalition issues are discussed further in Biggs and Smith (1998).
9. A recent review of institutionalisation experiences has stressed the need for organisational innovation (Ashby and Sperling 1995). In this chapter we have tried to look closely at the nature of this process and the way that it relates to the concept of social capital.

10. For a fuller explanation of this framework see Biggs and Matsaert (1999).
11. The significance of the agency role of specific development actors has been documented elsewhere (Biggs 1997; Jackson 1997; Goetz 1998) and is well illustrated in a recent integrated pest management (IPM) project. Soon after joining the project the social scientists wrote a report that enabled the whole team to reappraise the primary preoccupations of the project. Soil and water issues appeared to be far more important than IPM problems (Orr and Jere 1997).

References

Abbot, J. and I. Guijt. 1998. *Changing Views on Change: Participatory Approaches to Monitoring the Environment* (SARL Discussion Paper 2). London: International Institute for Environment and Development.

Abbot, J., E.N. Edmund and M.W. Khen. 1999. *Turning Our Eyes from the Forest (Report to BirdLife International)*. London: International Institute for Environment and Development.

Abdul, A.R. and P. Kerkhof. 1998. Forest monitoring by villagers. *Social Forestry & Environment* 4.

Adams, J.S. and T.O. McShane. 1992. *The Myth of Wild Africa: Conservation without Illusion*. New York and London: Norton.

Agrawal, B.D. & maize research team. 1979. *Maize On-farm Research Project (1979 Report)*. G. B. Pant University of Agriculture and Technology, Pantnagar, UP, India.

Amphlett, M.B. 1983. *Interim Report on the 1981/82 Rainy Season*. Soil Erosion Project, UK Hydraulics Research, Wallingford.

Anderson, D.M. 1984. Depression, dust bowl, demography and drought: The colonial state and soil conservation in East Africa during the 1930s. *African Affairs* 83, 232: 321–43.

Anderson, D. and R. Grove. 1987. *Conservation in Africa: People, Policies and Practice*. Cambridge: Cambridge University Press.

Anon. 1988. *Résultats de l'inventaire de la forêt classée de Takiéta pour le bois sec avec écorce*. PUSF.

Anon. 1989. *Forest Act 1989*. Khartoum: Republic of Sudan.

Anon. 1992. *Gestion de terroirs. Problèmes identifiés par les opérateurs de terrain en Afrique et en Madagascar. 1ᵉʳ document de synthèse (document provisoire)*: Reseau Recherche Développement.

Anon. 1993. *Principes d'orientation du code rural*. Niamey, Republique du Niger: Comité National du Code Rural.

Anon. 1995a. *Loi No. 95-003 portant organisation et l'exploitation du transport et du commerce du bois*. Bamako, Mali: Presidence de la République.

Anon. 1995b. *Loi No. 95-004 fixant les conditions de gestion des ressources forestières*. Bamako, Mali: Presidence de la République.

Anon. 1995c. Local level economic valuation of savanna woodland resources: village cases from Zimbabwe. *IIED Research Series* 3, 2.

Anon. 1995d. *The Hidden Harvest: The Value of Wild Resources in Agricultural Systems*. London.

Anon. 1997a. *Manuel de creation des marchés ruraux de bois*. Bamako, Mali: Cellule combustibles ligneux.

Anon. 1997b. *Rapport definitif de l'étude socio-économique dans la zone du Kelka*. Bamako, Mali: NEF.

Ashby, J.A. and L. Sperling. 1995. Institutionalising participatory, client-driven research and technology development in agriculture. *Development and Change* 26: 753–70.

Ayoo, S.J. 1995. *The Ik Research*. Kampala: OXFAM.

Ba, B. 1996. *Un conflit révélateur d'une crise de société: le cas du Leydi Soulali*. Sare-Mala.

Bacha, A.K. 1999. *Etudes des institutions nationales de gestion des forêts*. London: SOS Sahel (UK)/GDRN5.

Baker, P.R. 1974. *Perception of Pastoralism* (Discussion Paper no. 3). Norwich: School of Development Studies, University of East Anglia.

Baviskar, A. 1999. Participating in ecodevelopment: The case of the Great Himalayan National Park, Himachal Pradesh, pp. 109–29, in *A New Moral Economy for India's Forests?*, R. Jeffery and N. Sundar (eds). New Delhi: Sage.

Bayart, J.-F. 1993. *The State in Africa: The Politics of the Belly*. London and New York: Longman.

Behnke, R.H., I. Scoones and C. Kerven (eds) 1993. *Range Ecology at Disequilibrium: New Models of Natural Variability and Pastoral Adaptation in African Savannas*. London: ODI, IIED and Commonwealth Secretariat.

Bekule, S. *et al.* 1991. *Maasai Herding*. International Livestock Centre for Africa.

Belshaw, D., S. Avery, R. Hogg and R. Obin. 1996. *Report of the Evaluation Mission, Integrated Development in Karamoja, Uganda Project*. Norwich and New York: Overseas Development Group and UNCDF.

Berger, D.J. 1993. *Wildlife Extension: Participatory Conservation by the Maasai of Kenya*. Nairobi: ACTS Press.

Biggs, S.D. 1982. Generating agricultural technology: triticale for the Himalayan Hills. *Food Policy* 7, 1: 69–82.

Biggs, S.D. 1983. Monitoring and control in agricultural research systems: maize in Northern India. *Research Policy* 12: 37–59.

Biggs, S.D. 1997. Livelihood, coping and influencing strategies of rural development personnel. *Project Appraisal* 12: 101–6.

Biggs, S.D. and H. Matsaert. 1999. An actor oriented approach for strengthening research and development capabilities in natural resources systems. *Public Administration and Development* 19, 3: 231–62.

Biggs, S.D. and G. Smith. 1998. Beyond methodologies: Coalition building for participatory technology development. *World Development* 26: 239–48.

Birgegard, L.E. 1993. *Natural Resource Tenure: A Review of Issues and Experiences with Emphasis on Sub-Saharan Africa* (Rural Development Studies No. 3). Uppsala: IRDC, Swedish University of Agricultural Science.

Bishop, J. and I. Scoones. 1994. Beer and baskets: the economics of women's livel- ihoods in Ngamiland, Botswana. *IIED, The Hidden Harvest, Research Series* 3, 1. London.

BIT. 1997. *Projet aménagement des ressources forestières dans le Cercle de Kita. Rapport de la mission d'évaluation de fin de projet. Aide Mémoire*. Mali

Blaikie, P. 1985. *The Political Economy of Soil Erosion in Developing Countries*. Harlow: Longman.

Blaikie, P., K. Brown, M. Stocking, L. Tang, P. Dixon and P. Sillitoe. 1997. Knowledge in action: Local knowledge as a development resource and barriers to its incorporation in natural resource research and development. *Agricultural Systems* 55, 2: 217–38.

Bradbury, M., S. Fisher and C. Lane. 1995. *Working with Pastoralist NGOs and Land Conflict in Tanzania*. London: IIED.

Bromley, D.W. 1992. The commons, common property, and environmental policy. *Environmental and Resource Economics* 2, 1: 1–17.

Brown, D. 1998. Evaluating institutional sustainability in development programmes: Beyond dollars and cents. *Journal of International Development* 10, 1: 55–69.

Bunderson, W.T., F. Bodnar, W.A. Bromley and S.J. Nanthambwe. 1995. *A Field Manual for Agroforestry Practices in Malawi* (Malawi Agroforestry Extension Project, Publication Number 6). Malawi, Lilongwe: Malawi Agroforestry Extension Project.

Burnette, G.W. and K. Kang'ethe. 1994. Wilderness and the Bantu mind. *Environmental Ethics* 16.

Campbell, D. J. 1993. Land as ours, land as mine: Economic, political and ecological marginality in Kadiajo, pp. 258–72 in *Being Maasai: Ethnicity and Identity in East Africa*, T. Spear and R. Waller (eds). London: Currey.

Chakraborty, R.N., I. Freier, F. Kegel and M. Mäscher. 1997. *Community Forestry in the Terai Region of Nepal: Policy Issues, Experience, and Potential.* (Report No. 5). Berlin: German Development Institute.

Chambers, R. 1983. *Rural Development: Putting the Last First.* Harlow: Longman.

Chambers, R. 1997. *Whose Reality Counts? Putting the Last First.* London: Intermediate Technology Publications.

Chambers, R., A. Pacey and L.A. Thrupp. 1989. *Farmer First: Farmer Innovation and Agricultural Research.* London: Intermediate Technology Publications.

Chambers, R., N.C. Saxena and T. Shah. 1989. *To the Hands of the Poor: Water and Trees.* New Delhi: Oxford and IBH Publishing Company.

Chhetri, R.B. and M.C. Nurse. n.d. *Equity in User Group Forestry: Implementation of Community Forestry in Central Nepal.* Kathmandu: Nepal–Australia Community Forestry Project.

Child, B. 1996. The practice and principles of community-based wildlife management in Zimbabwe: the CAMPFIRE programme. *Biodiversity and Conservation* 5, 3: 369–98.

CIAT. 1996. *Plan estrat gico 1996–2001.* Santa Cruz, Bolivia: Centro de Investigación Agrícola Tropical.

Cisterino, M. 1979. Karamoja: The Human Zoo. PhD: University College of Wales, Swansea.

Collar, N.J. and S.N. Stuart. 1988. *Key Forests for Threatened Birds in Africa (Monograph 3).* Cambridge: ICBP.

Collett, G., R. Chhetri, W.J. Jackson and K.R. Shepherd. 1996. *Nepal–Australia Community Forestry Project: Socio-economic Impact Study.* Canberra: Anutech Pvt. Ltd. (Technical Note 1/96).

Cousins, B. 1996. Conflict Management for Multiple Resource Users in Pastoralist and Agro-pastoralist Contexts. Paper presented to the Third International Technical Consultation on Pastoral Development, 1996.

Cousins, C.C. and P. Reynolds. 1993. *Lwaano Lwanyika: Tonga Book of the Earth.* London: Panos.

DANIDA. 1993. *Agricultural Sector Review.* Copenhagen, Denmark: Danish International Development Agency.

Davies, P. 1994. *Bosquejo socioeconómico de Santa Cruz.* (Informe Técnico 16). Santa Cruz: Centro de Investigación Agrícola Tropical.

De Vries, P. 1992. A Research Journey, pp. 47–84 in *Battlefields of Knowledge,* N. Long and A. Long (eds). London: Routledge.

Devereux, S. 1997. *Household Food Security in Malawi* (Discussion Paper No. 362). University of Sussex: Institute of Development Studies.

Diakite, M. *Rapport de mission au projet de Kita*: Bamako, Mali: PPEB.

Dyson-Hudson. 1966. *Karimajong Politics*. Oxford: Clarendon Press.

Economic Review. 1997. Deadly situation: Drought leaves pastoralists and livestock starving, in *Economic Review*. Nairobi.

Ellis, J.E. and D.M. Swift. 1988. Stability of African pastoral ecosystems: Alternative paradigms and implications for development. *Journal of Range Management* 41: 450–9.

Elrick, D.E. and W.D. Reynolds. 1992. Methods for analysing constant-head well permeameter data. *Soil Science Society of America Journal* 56: 320–3.

Fairhead, J. and M. Leach. 1994. Declarations of difference, pp. 75–9 in *Beyond Farmer First*, I. Scoones and J. Thompson (eds). London: Intermediate Technology Publications.

Farrington, J. 1998. Organisational roles in farmer participatory research and extension: lessons from the last decade. *ODI Natural Resource Perspectives* 27, 1: 1–10.

Feldhaus, A. 1995. *Water and Womanhood: Religious Meanings of Rivers in Maharashtra*. New York: Oxford University Press.

Fiallo, E.A. and S.K. Jacobsen. 1995. Local communities and protected areas: Attitudes of rural residents towards conservation and Machalilla National Park, Ecuador. *Environmental Conservation* 22, 3: 241–9.

Fisher, R.J. 1995. *Collaborative Management of Forests for Conservation and Development*. Gland, Switzerland and Cambridge: IUCN.

Fleissner, K. 1996. Bambara Groundnut Improvement Programme, pp. 1.38–1.42, *1995/96 Progress Report*. Windhoek: Government of Namibia, Ministry of Agriculture, Water and Rural Development Division of Plant Production Research.

Foley, G. *et al.* 1997. *The Niger Household Energy Project* (World Bank Technical Paper). Washington DC: World Bank.

Food and Agriculture Organisation. 1991. A study of the reasons for the success or failure of soil conservation projects. *FAO Soils Bulletin* 64.

Freire, P. 1970. *Pedagogy of the Oppressed*. New York: Herder and Herder.

Freire, P. 1973. *Education for Critical Consciousness*. New York: Seabury Press.

Fujisaka, S. 1994. Learning from six reasons why farmers do not adopt innovations intended to improve sustainability of upland agriculture. *Agricultural Systems* 46, 4: 409–25.

Galaty, J. 1988. The Maasai group-ranch: Politics and development in an African pastoral society, in *When Nomads Settle*, P.C. Salzman (ed.). New York: J. F. Bergin.

Gardner, A., S. Anders, C. Asanga and J. Nkengla. 1997. Community forest management at the Kilum–Ijim Forest Project: Lessons learned so far, pp. 62–7, in *African Rainforests and the Conservation of Biodiversity*, S. Doolan (ed.). Oxford: Earthwatch Europe.

Gardner, R. 1997. The third dimension: Soil fertility and integrated nutrient management on hillsides, in *Integrated Nutrient Management on Farmers' Fields: Approaches that Work* (Occasional Publication No. 1), P.J. Gregory, C.J. Pilbeam and S.H. Walker (eds). Reading: The Department of Soil Science, The University of Reading.

Gazette. 1992. In *The Gazette*. Montreal, Southam Inc.

Giraud, S. 1998. *Les aménagements villageois du massif de Tientiergou: bilan socio-technique après six ans de fonctionnement*.: Projet Energie-II/CIRAD Foret-FIF-ENGREF.

Goetz, A.M. 1998. Local heroes: Patterns of field worker discretion in implementing GAD policy in Bangladesh. Paper presented to the Proceedings of a Workshop on Recent Research on Micro-Finance: Implications for Policy, Rural Research Unit at Sussex Working Papers No. 3, 1998.

Goldsmith, A.A. 1993. Institutional development in national agricultural research: issues for impact assessment. *Public Administration and Development* 13, 3: 195–204.

Gordon, H.S. 1954. The economic theory of a common-property resource: The fishery. *Journal of Political Economy* 62: 124–42.

Government of Zimbabwe. 1984. *Prime Minister's Directive on Ward Development Committees and Village Development Committees*. Harare: Government Printers.

Government of Zimbabwe. 1992. *Census 1992 Zimbabwe Preliminary Report*. Harare: Statistical Office, Government Printers.

Graham, D. 1995. Kenya wildlife chief proud of new harmony with villagers, in *The San Diego Union Tribune*.

Graham, O. 1989. A land divided: the impact of ranching on a pastoral society. *The Ecologist* 19, 5: 184–5.

Graner, E. 1997. *The Political Ecology of Community Forestry in Nepal*. Saarbrücken: Verlag für Entwicklungspolitik.

Grindle, M.S. 1997. Divergent cultures? When public organisations perform well in developing countries. *World Development* 25, 4: 481–96.

Gulliver, P.H. 1970. *The Family Herds: A Study of Two Pastoral Tribes in East Africa, the Jie and Turkana*. Westport: Negro University Press.

Gupta, A. 1998. *Postcolonial Development: Agriculture in the Making of Modern India*. Durham, NC and London: Duke University Press.

Hardin, G. 1968. The tragedy of the commons. *Science* 162: 1243–8.

Harkes, I. 2001. Project success: different perspectives, different measurements. In *Analytical Issues in Participatory Natural Resource Management*, B. Vira and R. Jeffery (eds). London: Palgrave.

Harriss, J. and P.D. Renzio. 1997. Policy Arena, 'Missing Link' or Analytically Missing? The Concept of Social Capital. *Journal of International Development* 9, 7: 919–37.

Hasler, R. 1996. *Agricultural Foraging and Wildlife Use in Africa*. London and New York: Kegan Paul International.

Heinen, J.T. 1993. Park-people relations in Kosi Tappu Wildlife Reserve, Nepal: A socio-economic analysis. *Environmental Conservation* 20, 1: 25–34.

Heverkort, B., W. Hiemstra, D. Millar and W. Rist. 1996. *Experimenting with Farmers' Worldview*. (Platform for Intercultural Dialogue and Cosmovision). Amsterdam: COMPAS Luesden.

Hillman, E. 1994. The pauperization of the Maasai in Kenya. *Africa Today* 41, 4: 57–65.

His Majesty's Government and FINNIDA. 1993. *Forest Resources of the Terai Districts 1990/91* (Publication No. 57). Kathmandu: Forest Research and Survey Centre (Ministry of Forests and Soil Conservation) and Forestry Sector Institutional Strengthening Programme (FINNIDA).

His Majesty's Government and FINNIDA. 1994. *Deforestation in the Terai districts 1978/79–1990/91* (Publication No. 60). Kathmandu: Forest Research and Survey Centre (Ministry of Forests and Soil Conservation) and Forest Resource Information System Project (FINNIDA).

Hobley, M. 1996. *Participatory Forestry: The Process of Change in India and Nepal.* London: Overseas Development Institute.

Hopkins, C. 1992. *1992 Remeasurement of the 1982–83 Test Cut at the Guesselbodi Forest*: Projet Energie 2, Volet Offre.

Hudson, N.W. 1987. Soil and water conservation in semi-arid areas. *FAO Soils Bulletin 57.*

Hudson, N. 1995. *Soil Conservation.* London: BT Batsford.

Human Rights Watch/Africa Watch. 1993. *Divide and Rule: State-Sponsored Ethnic Violence in Kenya.* New York: Human Rights Watch.

Hutter, R. and O. Fulling. 1994. *Mammal Diversity in the Oku Mountains, Cameroon*: Bonn Museum.

IFAD. 1986. *Soil and Water Conservation in Africa: Issues and Options.* Amsterdam: Free University of Amsterdam. Centre for Development Co-operation Services.

IIED. 1994. *Whose Eden? An Overview of Community Approaches to Wildlife Management.* London: International Institute for Environment and Development.

Infield, M. 1989. Hunters claim a stake in the forest. *New Scientist*, 4 November: 52–5.

International Council for Bird Preservation. 1992. *Putting Biodiversity on the Map: Priority Areas for Global Conservation.* Cambridge: ICBP.

ISNAR. 1993. *Review of the Agricultural Research System in Namibia.* ISNAR 56.

Jackson, C. 1997. Sustainable development at the sharp end: Fieldworker agency in a participatory project. *Development in Practice* 7, 3: 237–47.

Jackson, W.J. and A.W. Ingles. 1998. *Participatory Techniques for Community Forestry: A Field Manual.* Gland, Switzerland and Cambridge: IUCN and WorldWide Fund for Nature.

Jeffery, R. and N. Sundar (eds) 1999. *A New Moral Economy for India's Forests?* New Delhi, London and Newbury Park: Sage.

Jeffery, R., N. Sundar and P. Khanna. 2001. Joint forest management: A silent revolution amongst forest department staff?, in B. Vira and R. Jeffery (eds) *Analytic Issues in Participatory Natural Resource Management.* Basingstoke: Palgrave.

Joel, D.R. and P.J. Sanjeeva Raj. 1988. Distribution and zonation of shore crabs in the Pulicat Lake. *Proceedings of the Indian Academy of Sciences (Animal Sciences)* 95, 4: 437–45.

Kaliyamurthy, M, 1972. Observations on the food and feeding habits of some fishes of the Pulicat Lake, *Journal of the Inland Fish Society. India* 4: 115–21.

Karengata, C. 1997. *Maasai on the Warpath.* Nairobi, Kenya.

Karki, M., J.B.S. Karki and N. Karki. 1994. *Sustainable Management of Common Forest Resources: An Evaluation of Selected Forest User Groups in Western Nepal (Case Studies of Palpa District and the Phewa Watershed).* Kathmandu: International Centre for Integrated Mountain Development (ICIMOD).

Kellert, S.R. 1996. *The Value of Life.* Washington DC: Island Press.

Kenmuir, R. 1978. *A Wilderness Called Kariba: The Wildlife and Natural History of Lake Kariba.* Harare: Mardon Printers.

Kerkhof, P. 1990. *Agroforestry in Africa. A Survey of Project Experience.* London: Panos Institute.

Kerkhof, P. 1996. *Mali Field Report.* London: SOS Sahel (UK).

Kerkhof, P. 1997a. *Mali Field Report.* London: SOS Sahel (UK).

Kerkhof, P. 1997b. *Sudan Field Report.* London: SOS Sahel (UK).

Kerkhof, P. 1998. *Mali Field Report.* London: SOS Sahel (UK).

Kerkhof, P. 1999. *Sudan Field Report.* London: SOS Sahel (UK).

Kerr, J.M. and N.K. Sanghi. 1992. *Soil and Water Conservation in India's Semi Arid Tropics* (Sustainable Agriculture Programme Gatekeeper Series SA34). London: International Institute for Environment and Development.

Kipury, N. 1983. *Oral Literature of the Maasai.* Nairobi: Heinemann Educational Books.

Kituyi, M. 1990. *Becoming Kenyans: Socio-economic Transformation of the Pastoral Maasai.* Nairobi: ACTS.

Konate, A.B. and M.M. Tessougue. 1996. *La gestion des ressources naturelles renouvelables dans la forêt du Samori.* London: SOS Sahel (UK).

Kothari, A., N. Pathak, R.V. Anuradha and B. Taneja (eds) 1998. *Communities and Conservation: Natural Resource Management in South and Central Asia.* New Delhi: Sage.

KPIU. 1996. *Karamoja Project Implementation Unit: Inception Report (Draft).* Kampala: Karamoja Development Agency, Ministry of State for Karamoja.

Lamphear, J. 1976. *The Traditional History of the Jie of Uganda.* Oxford: Clarendon Press.

Lane, J. 1995. Non-governmental organisations and participatory development: the concept in theory versus the concept in practice. In *Power and Participatory Development: Theory and Practice*, N. Nelson and S. Wright (eds). London: Intermediate Technology Publications.

Laurent, P.J. and P. Mathieu. n.d. *Migration, environnement et projet de développement. Récit d'un conflit foncier entre Nunni et Mossi au Burkina Fasso.* Louvain la Neuve: Institut d'Etudes de developpement.

Lawrence, A., G. Haylor, C. Barahona and E. Meusch. In press. Adapting participatory methods to meet different stakeholder needs: Farmers' experiments in Bolivia and Laos, in *Learning from Change: Issues and Challenges in Participatory Monitoring and Evaluation* M. Estrella, J. Blauert and J. Gaventa (eds). London: Intermediate Technology Publications.

Le Roy, E. *et al.* 1996. *La securisation foncière en Afrique. Pour une gestion viable des ressources renouvelables*: Karthala.

Leach, M. and J. Kamangira. 1996. *PAPPPA Benficiary Adoption and Assessment Survey 1996.* Lilongwe, Malawi: PAPPPA Management Unit.

Leach, M. and N. Marsland. 1994. *ADDFOOD Beneficiary Assessment Study 1994 using Participatory Rural Appraisal methods.* Lilongwe, Malawi: Planning Division, Ministry of Agriculture and Livestock Development.

Leach, M. and R. Mearns (eds) 1996. *The Lie of the Land: Challenging Received Wisdom on the African Environment.* Oxford: James Currey and International African Institute.

Leach, M., R. Mearns and I. Scoones. 1997. Challenges to community-based sustainable development. *IDS Bulletin* 28, 4: 4–14.

Lindblade, K. 1996. Reminiscences of Uganda, in *Newsletter of the Population-Environment Fellows Program,* K. Lindblade (ed.). University of Michigan.

Lisa, G. *et al.* 1996. From protection to participation. In K. Lindblade (ed.) *Newsletter of the Population-Environment Fellows Program.* University of Michigan.

Long, N. and M. Villareal. 1994. The interweaving of knowledge and power in development interfaces, pp. 41–51, in *Beyond Farmer First*, I. Scoones and J. Thompson (eds). London: Intermediate Technology Publications.

Lund, C. 1995. *Law, Power and Politics in Niger.* Roskilde University: International Development Studies.

Madougou, J. 1999a. *Etude socio-economique du village de Tientiergou après six ans de gestion locale de la forêt villageoise, Arrondissement de Say, Niger.* London: SOS Sahel (UK).

Madougou, J. 1999b. *Sociologie des agents forestiers: Cas du Niger.* London: SOS Sahel (UK).

Mahila Upakar Community Forestry User Group. 1996. *Constitution.* Kohalpur, Nepal: mimeo.

Maiga, I. and G.S.A. Diallo. 1995. *Recherche sur les problèmes fonciers au Mali: études de cas des litiges dans la région de Mopti*: London: GRAD/IIED.

Maisels, F.G. 1999. *GIS and Map Work, Progress Report.* London: BirdLife International Kilum–Ijim Forest Project: mimeo.

Mamdani, M., P.M.B. Kasoma and A.B. Latende. 1992. *Karamoja: Ecology and History* (Centre for Basic Research Working Paper no. 22). Kampala: CBR Publications.

Matampash, K. 1993. The Maasai of Kenya. In *Indigenous Views of Land and the Environment*, S. Davis (ed.). Washington DC: World Bank.

Mathew, S. 1991. *Study of Territorial Use Rights in Small Scale Fisheries: Traditional Systems of Fisheries Management in Pulicat Lake, Tamil Nadu, India* (FAO Fisheries Circular 839). Rome: Food and Agriculture Organisation.

Mathias-Mundy, E., G. Morin-Labatut and S. Akhtar. 1993. Background to the International Symposium on Indigenous Knowledge and Sustainable Development. *Indigenous Knowledge and Development Monitor.*

Maveneke, T.N. 1992. The Role of Wildlife in Traditional Customs and Benefits to Local Communities of Wildlife Utilisation. CAMPFIRE Zimbabwe. Paper presented to the Wildlife Producers Association of Swaziland, 1992.

McKinley, J. 1996. Warily, the Maasai embrace the animal kingdom. In *The New York Times.*

Mellis, D., H. Skinner and M. Boniface. 1997. Tillage research challenges tool-makers in Kenya, pp. 127–38, in *Farmers' Research in Practice: Lessons from the Field*, L.V. Veldhuizen, A.R. Water-Bayer, D.A. Johnson and J. Thompson (eds). London: IT Publications.

Ministry of Agriculture. 1993. *ADDFOOD Beneficiary Assessment Survey 1993 Report.* Lilongwe, Malawi: Planning Division, Ministry of Agriculture.

Ministry of Agriculture. 1995a. *Agricultural Handbook 1995.* Lilongwe, Malawi: Ministry of Agriculture.

Ministry of Agriculture. 1995b. *Financing Proposal for the Years 1996–2001 of the Project 'Promotion of Soil Conservation and Rural Production'.* Lilongwe, Malawi: Ministry of Agriculture and PROSCARP.

Ministry of the Environment and Forestry. 1997. *Manual of the Procedures for the Attribution, and Norms for the Management, of Community Forests.* Yaoundé, Cameroon: MINEF.

Mitchell, F. 1969. *Forecasts of Returns to Kajiado County Council from Maasai Amboseli Game Reserves, 1970–2000* (Institute of Development Studies, Paper 87). Nairobi: University of Nairobi.

Mkanda, F.X. and S.M. Munthali. 1994. Public attitudes and needs around Kasungu National Park, Malawi. *Biodiversity Conservation* 3, 1: 29–44.

Moss, J.M.S. 1996. *The Regeneration Dynamics of Arid* Acacia tortilis *Woodland Formations*. Northern Kenya: OFI.

Mounkaila, M. 1997. *Inventaire des ressources forestières ligneuses de la forêt classée de Takiéta*. Institut Developpement Rural Kollo.

Müller-Böker, U. 1995. *Die Tharu in Chitawan. Kenntnis, Bewertung und Nutzung der natürlichen Umwelt im südlichen Nepal*. Stuttgart: Steiner.

Murphree, M.W. 1991. *Communities as Institutions for Resource Management*. Harare: CASS University of Zimbabwe.

Murphree, M.W. 1993. *Communities as Resource Management Institutions Sustainable Agriculture* (Gatekeepers series No. 36). London: International Institute for Environment and Development.

Nadkarni, M.V., S.A. Pasha and L.S. Prabhakar. 1989. *The Political Economy of Forest Use and Management*. New Delhi, Newbury Park and London: Sage.

Narayan, D. 1993. Participatory evaluation: tools for managing change in water and sanitation. Washington DC: The World Bank.

Nation Group of Newspapers. 1994a. Leakey must go, say Ntimama; and why Dr. Leakey might never know what hit him. In *Nation Group of Newspapers*, 16 January.

Nation Group of Newspapers. 1994b. Leakey gains support. In *Nation Group of Newspapers*, 11 January.

Nation Group of Newspapers. 1994c. Team cites irregular procedures; and What the committee established. In *Nation Group of Newspapers*, 2 April.

Nation Group of Newspapers. 1994d. Embattled Leakey quits as KWS boss. In *Nation Group of Newspapers*, 15 January.

Nation Group of Newspapers. 1994e. Moi rejects quit letter by leakey. In *Nation Group of Newspapers*, 11 March.

Nation Group of Newspapers. 1994f. Poaching: Government challenged. In *Nation Group of Newspapers*, 13 March.

Nation Group of Newspapers. 1994g. Leakey rejects terms and quits. In *Nation Group of Newspapers*, 24 March.

Nation Group of Newspapers. 1996. Fight for Maasai rights. In *Nation Group of Newspapers*, 28 November.

Nation Group of Newspapers. 1997a. Morans 'raid' city school. In *Nation Group of Newspapers*, 23 September.

Nation Group of Newspapers. 1997b. Campaign to lift the ban on hunting: What is at stake? In *Nation Group of Newspapers*, 23 March.

Nation Group of Newspapers. 1997c. Why are KWS donors pulling out? In *Nation Group of Newspapers*, 9 November.

Nation Group of Newspapers. 1997d. Hunting saga: the perils that stalk wildlife. In *Nation Group of Newspapers*, 30 March.

Nation Group of Newspapers. 1997e. Step up conservation efforts. In *Nation Group of Newspapers*, 19 December.

Nation Group of Newspapers. 1997f. MP issue ethnic warning: Kipsigis won't rule Narok-Tuya. In *Nation Group of Newspapers*, 25 August.

Nation Group of Newspapers. 1997g. Andrew Ngwiri. Is time running out? In *Nation Group of Newspapers*, 2 February.

Nelson, N. and S. Wright (eds) 1995. *Power and Participatory Development: Theory and Practice*. London: Intermediate Technology Publications.

Nelson, N. and S. Wright. 1995. Participation and power, in *Power and Participatory Development: Theory and Practice*, N. Nelson and S. Wright (eds). London: Intermediate Technology Publications.

Nepal, S.K. and K.E. Weber. 1995. Prospects for co-existence: wildlife and local people. *Ambio* 24, 4: 238–45.

Nessler, U. 1980. Ancient techniques aid modern arid zone agriculture. *Israel Journal of Development* 20, 5: 3–7.

Ngoleka, J.R.M. n.d. History of Soil Conservation in Malawi. Paper presented at the Southern African Development Co-ordination Conference. Zomba, Malawi: mimeo.

Niamir, M. 1997. The resilience of pastoral herding in Sahelian Africa. In *Linking Social and Ecological Systems: Institutional Learning for Resilience*, F. Berkes and C. Folke (eds). Cambridge: Cambridge University Press.

Novelli, B. 1988. *Aspects of Karimajong Ethnosociology* (Museum Combonianum no. 44). Kampala: Comboni Missionaries.

Noy-Meir and B.H. Walker. Stability and resilience in Rangelands, pp. 21–5, in *Dynamics of Range Ecosystems*.

Ocan, C.E. 1992. *Pastoral Crisis in Northeast Uganda: The Changing Significance of Cattle Raids* (Centre for Basic Research, Working Paper no. 21). Kampala: CBR.

Okali, C., J. Sumberg and F. Farrington. 1994. *Farmer Participatory Research: Rhetoric and Reality*. London: IT Publications.

Okondo, P. 1993. The strengths of Majimboism. *The Economic Review*.

Orr, A. and P. Jere. 1997. *What Have We Learnt? 1996/97 in Review*. Limbe, Malawi: Farming Systems Integrated Pest Management Project, Ministry of Agriculture and Livestock development, Department of Agricultural Research, Farming Systems, Bvumbwe Research Station.

Ostrom, E. 1990. *Governing the Commons: The Evolution of Institutions for Collective Action*. Cambridge: Cambridge University Press.

Ostrom, E. 1996. Crossing the great divide: Coproduction, synergy and development. *World Development* 24, 6: 1073–87.

Ostrom, E. 1999. *Self-Governance and Forest Resource* (CIFOR Occasional Paper 20). Bogor, Indonesia: Center for International Forestry Research.

Ostrom, E., R. Gardner and J. Walker. 1994. *Rules, Games, and Common-Pool Resources*. Ann Arbor: University of Michigan Press.

Painter, M. 1995. Upland–lowland production linkages and land degradation in Bolivia, pp. 133–68 in *The Social Causes of Environmental Destruction in Latin America*, M. Painter and W.H. Durham (eds). Ann Arbor, Michigan: University of Michigan Press.

Paris, P. 1997. *Mission de consultation pour l'élaboration d'un cadre thématique pour le plan d'aménagement pastoral de la zone péripherique de Baban Rafi*. CARE.

Parry, D. and B. Campbell. 1992. Attitudes of rural communities to animal wildlife and its utilization in Chobe Enclave and Mababe Depression, Botswana. *Environmental Conservation* 19, 3: 245–52.

Pazzaglia, A. 1982. *The Karimajong: Some Aspects*. Bologna, Italy: EMI Press.

Pearce, F. 1998. Ecological vandals: While the elephants are the No. 1 tourist attraction in Kenya's parks and reserves, they also are capable of quickly turning thick bush into desert. In *Pittsburgh Post-Gazette*.

Peltier, R. (ed.) 1994. *Aménagement villagois du massif de brousse tachetée de Tientiergou* (Rapport technique no. 32). Niamey, Niger: Projet Energie II – Energie Domestique.

Posey, D.A. (ed.) 1998. *Cultural and Spiritual Values of Biodiversity: Global Biodiversity Assessment*. Oxford: Oxford Centre for Environment, Ethics and Society, A UNEP Publication.

Pretty, J.N. 1994. Alternative systems of inquiry for a sustainable agriculture. *IDS Bulletin* 25, 2: 37–48.

Pretty, J.N. 1995. *Regenerating Agriculture: Policies and Practice for Sustainability and Self-reliance*. London: Earthscan.

Pretty, J.N. and P. Shah. 1994. *Soil and Water Conservation in the Twentieth Century: A History of Coercion and Control* (Research series No. 1). University of Reading.

Rao, J.M. and J.M. Caballero. 1990. Agricultural performance and development strategy: Retrospect and prospect. *World Development* 18, 6: 899–913.

Rao, P. and K. K. Mohapatra. 1993. Wetland Avifauna of Pulicat bird sanctuary, pp. 11–15 in *Proceedings of the 1st Seminar on Changing Scenario of Bird Ecology and Conservation*. Bangalore: Ornithological Society of India.

Rattray, J.M. and F.O. Byrne. 1963. *Report to the Government of Uganda on a Reconnaissance Survey of the Karamoja District, Project no. UGA/TE/AN*. FAO.

Reij, C., S.D. Turner and T. Kuhlman. 1986. *Soil and Water Conservation in Sub-Saharan Africa: Issues and Options*. Rome: IFAD.

Republic of Kenya. 1989. The 1989 Population Census. Nairobi: Government Printer.

Republic of Kenya. 1990. *Kajiado District Atlas*. Kajiado District, Kenya: ASAL Program.

Republic of Kenya. 1994. *Kajiado District Development Plan 1994-96*. Nairobi: Government Printer.

Republic of Kenya. 1997. *8th Development Plan*. Nairobi: Government Printer.

Richard, W. 1988. Emutai: Crisis and response in Maasailand 1883–1902. In *The Ecology of Survival: Case Studies from Northeast African History*, D.H. Johnson and D.M. Anderson (eds). Boulder, Colorado: Westview Press.

Richard, W. 1993. Acceptees and aliens: Kikuyu settlement in Maasailand. In *Being Maasai: Ethnicity and Identity in East Africa*, T. Spear (ed.). London: James Currey.

Richards, M. 1994. Towards valuation of forest conservation benefits in developing countries. *Environmental Conservation* 21, 4: 308–19.

Roling, N. 1996. Creating human platforms to manage natural resources: First results of a research programme, pp. 51–64 in *Agricultural RandD at the Crossroads: Merging Systems Research and Social Actor Approaches*, A. Budelman. (ed.) Amsterdam: Royal Tropical Institute.

Ross, M. 1995. Interests and identities in natural resource conflicts involving indigenous peoples. *Cultural Survival Quarterly* 19, 3.

Rugege, S. 1995. Conflict resolution in African Customary Law. *Africa Notes*, October.

Rutten, M.M.E.M. 1992. *Selling Wealth to Buy Poverty: The Process of the Individualisation of Landownership Among the Maasai Pastoralists of Kajiado District, Kenya, 1890–1990.* Fort Lauderdale: Verleg Breitenbach Publishers.

Sani, A.M. 1996. *Etude preliminaire sur le cadre institutionel et juridique des comités locaux de gestion des ressources naturelles.* SDSA II/UICN.

Sankan, S.S. 1971. *The Maasai.* Nairobi: East African Literature Bureau.

Savyasaachi. 1999. Sapangada: A Kuianka living space, pp. 130–50, in *A New Moral Economy for India's Forests?,* R. Jeffery and N. Sundar (eds). New Delhi, London and Newbury Park: Sage.

Saxena, N.C. and M. Sarin. 1999. Western Ghats Forestry Project in Karnataka: A preliminary assessment, pp. 181–215 in *A New Moral Economy for India's Forests?* R. Jeffery and N. Sundar (eds). New Delhi: Sage.

Scoones, I. 1997. The dynamics of soil fertility change: Historical perspectives on environmental transformation from Zimbabwe. *Geographical Journal* 163, 2: 161–9.

Shepherd, G. and D. Brown. 1997. Linking international priority setting to local institutional management, pp. 77–87 in *African Rainforests and the Conservation of Biodiversity,* S. Doolan (ed.). Oxford: Earthwatch Europe.

Shiva, V. 1991. *Ecology and the Politics of Survival.* New Delhi: Sage Publications.

Sibanda, B.M.C. 1995. Wildlife conservation and the Tonga in Omay. *Land Use Policy* 12, 1: 69–85.

Sibanda, B.M.C. 1996. Omay wildlife: Local institutions are the key. *Environmental Policy and Practice* 6, 1: 33–40.

Sibanda, B.M.C. 1997. Governance and environment: The role of African religion in sustainable utilisation of natural resources in Zimbabwe. *Forests, Trees and People Newsletter* 16, 34: 27–31.

Sibanda, B.M.C. 1998. Community based natural resources management systems: An evaluation of the CAMPFIRE programme in Zimbabwe: with special reference to Omay and Makande Communal Areas in Nyaminyami District. PhD: Rhodes University.

Sims, B.G., F. Rodríguez, M. Eid and T. Espinoza. 1999. Biophysical aspects of vegetative soil and water conservation practices in the Inter-Andean Valleys of Bolivia. *Mountain Research and Development* 19, 4.

Sindiga, I. 1981. European perceptions as a factor in the degrading Maasai ecology. M.A. Thesis: Ohio University,.

Sohani, G.G., *et al.* 1998. *Conjunctive Use of Water Resources in Deccan Trap, India.* Pune: BAIF Development Research Foundation.

Soussan, J., E. Gevers, K. Ghimire and P. O'Keefe. 1991. Planning for sustainability: access to fuelwood in Dhanusha District, Nepal. *World Development* 19, 10: 1299–314.

Srivastava, J.P. 1973. Prospects for triticale as a commercial crop in India. Paper presented to the Triticale International Symposium, El Batan, Mexico, 1973.

Stattersfield, A.J., M.J. Crosby, A.J. Long and D.C. Wege (eds) 1998. *Endemic Bird Areas of the World: Priorities for Biodiversity Conservation.* Cambridge: BirdLife International.

Stocking, M. 1985. Soil Conservation Policy in Colonial Africa. *Agricultural History* 59: 148–61.

Stocking, M. 1996. Breaking New Ground, pp. 140–54 in *The Lie of the Land: Challenging Received Wisdom on the African Environment* M. Leach and R. Mearns (eds). Oxford: International African Institute and James Currey.

Stuart, S.N., R.J. Adams and M.D. Jenkins (eds) 1990. *Biodiversity in sub-Saharan Africa and its Islands*. Gland, Switzerland: IUCN.

Subbarao, K.V. and P.R. Hooper. 1988. Reconnaissance Geological Map of the Deccan Basalt Group in the Western Ghats, India. In *Deccan Flood Basalts (Memoir 10)* (ed.) K.V. Subbarao. Bangalore: Geological Society of India.

Sunder Raj, S.K. and Sanjeeva Raj, P.J. 1987. Polychaeta of the Pulicat Lake. *Journal of the Bombay Natural History Society.* 84, 1: 84–104.

Sutherland, A., A. Martin and J. Salmon. 1998. *Recent Experiences with Participatory Technology Development in Africa: Practitioners' View*, Natural Resource Perspectives No. 25. London: ODI.

Sutherland, A., *et al.* 1995. Supply or lie: Linking research and sustainable technology supply through participatory technology development in dryland farming areas. Paper presented to the 5th KARI Biennial Scientific Conference, Kenya, 1995.

Sylla, M. 1996. *Etude d'inventaire du massif de Kelka*. Mali: BICOF/NEF.

Tagare, G.V. 1992. Water exploration in ancient India. *The Deccan Geographer* 30, 43–8.

Talbott, K. and S. Khadka. 1994. *'Handing it Over': An Analysis of the Legal and Policy Framework of Community Forestry in Nepal*. Washington DC: World Resources Institute.

Tendler, J. 1997. *Good Government in the Tropics*. Baltimore: The Johns Hopkins University Press.

Thangavelu, R. and Sanjeeva Raj, P.J. 1988. 'Distribution of molluscan fauna in Pulicat Lake', National Seminar on Shellfish Research and Farming, *CMFRI Bulletin*. 42, 1: 58–67.

The Guardian. 1999. Environment: Game plan. In *The Guardian*, 28 April.

The People. 1997. The gravy train that's KWS. In *The People*. 31 October–6 November.

Thomas, D.W. 1986. Vegetation in the Montane Forest of Cameroon, pp. 20–7 in *The Conservation of the Montane Forests of Western Cameroon (Report)*, S.N. Stuart (ed.). Cambridge: International Council for Bird Preservation.

Thomas, D.W. 1987. Vegetation of Mount Oku, pp. 54–6 in *Conservation of Oku Mountain Forest, Cameroon*, H. Macleod (ed.). ICBP Report No. 15. Cambridge: International Council for Bird Preservation.

Thomas, S.J. 1991. *The Legacy of Dualism and Decision-Making*. Harare: CASS University of Zimbabwe Press.

Tiffen, M., M. Mortimore and F. Gichuki. 1994. *More People, Less Erosion: Environmental Recovery in Kenya*. Colchester: ODI and Wiley.

Tignor, R.L. 1976. *The Colonial Transformation of Kenya*. Princeton: Princeton University Press.

Tremmel, M. 1996. *The People of the Great River* (Silveira House Series No. 9). Harare, Zimbabwe: Mambo Press.

Tropp, H. 1999. The role of voluntary organisations in environmental service provision, pp. 113–45 in *State, Society and the Environment in South Asia*, S.T. Madsen (ed.). London: Curzon.

Turton, C., A. Vaidya, K.D. Joshi and J.K. Tuladhar. 1995. *Towards Fertility Management in the Hills of Nepal*. Chatham: Natural Resources Institute.

UK Government. 1997. *Eliminating World Poverty: A Challenge for the 21st Century*. London: HMSO Publications.

UNEP. 1997. *Cultural and Spiritual Values of Biodiversity* (draft).

United Nations and the Government of Malawi. 1993. *Situational Analysis of Poverty in Malawi*. Limbe, Malawi: Montford Press.

Vargas, M.J.O., A. Kress, O.P.G. Montesinos, C.B. Mercado, C.R. Arquipino and E. Silva. 1993. *Plan Maestro Forestal, Provincia de Vallegrande*. Cochabamba: Comité Interinstitucional de Recursos Naturals de Vallegrande.

Vigdis, B.-D. *et al.* 1980. Women and Pastoral Development: Some Research Priorities for the Social Sciences. Paper presented to the Workshop on the Future of Pastoral Peoples, Kenya, 4–8 August.

Vira, B. 1999. Implementing JFM in the field: Towards an understanding of the Community-bureaucracy Interface, pp. 254–75 in *A New Moral Economy for India's Forests?*, R. Jeffery and N. Sundar (eds). New Delhi: Sage.

Water and Energy Commission Secretariat. 1996. *Energy Synopsis Report: Nepal 1994/95*. His Majesty's Government of Nepal, Ministry of Water Resources.

Weekly Review. 1997. Nothing to worry about. In *Weekly Review*, Nairobi, 7 November.

Western, D. 1969. *Land Use in Maasai Amboseli Game Reserve: A Case-study for Inter-Disciplinary Discussion*. (IDS paper no. 40). Nairobi: University of Nairobi.

Western, D. 1994. Ecosystem conservation and rural development: the case of Amboseli. In *Natural Connections: Perspectives in Community Based Organization*, D. Western and M. Wright (eds). Washington, DC: Island Press.

White, F. 1981. The history of the Afromontane archipelago and the scientific need for its conservation. *African Journal of Ecology* 19, 1: 33–54.

Wild, C.J. 1994. The status and ecology of the montane Herpetofaun of Mt Oku, Cameroon, Africa, pp. 73–91 in *Proceedings of the 1994 Symposium on Ecology and Conservation of Reptiles and Amphibians*, D. Reason (ed.). Oxford: Association for the Study of Reptilia-Amphibia.

Williams, D. and T. Young. 1994. Governance, the World Bank and liberal theory. *Political Studies* XLII, 1: 84–100.

Winpenny, J.T. 1993. *Values for the Environment. A Guide to Economic Appraisal*: ODI/HMSO.

Wood, G.D. 1994. *Bangladesh: Whose Ideas? Whose Interests?* Dhaka: University Press.

World Bank. 1992. *Economic Report on Environmental Policy* (Vols. 1 and 2). Lilongwe, Malawi: World Bank.

World Bank. 1995. *World Bank Sourcebook on Participation*. Washington DC: Environment Department, The World Bank.

Wunder, M. 1996. IUCN World Conservation Congress and Co-Management. In *Newsletter of the Population–Environment Fellows Program*, K. Lindblade (ed.). University of Michigan.

Yadama, G. and M. DeWeese Boyd. 2001. Co-management of forests in the tribal regions of Andhra Pradesh, India: a study in the making and unmaking of social capital. In *Analytical Issues in Participatory Natural Resource Management*, B. Vira and R. Jeffery (eds). London: Palgrave.

Young, A. 1989. *Agroforestry for Soil Conservation*. Wallingford, Oxon: CAB International.

Zimbabwe Trust. 1989. *Zimbabwe's Alternative to Ivory Ban*. Harare: Zimbabwe Trust.

Index

ADD 'food for work' (ADDFOOD), 80, 82
Agricultural Research Institute, 217
agrochemicals, 173
Argentina, 178

Bangladesh, 224
Bay of Bengal, 65, 71
Biggs, S., 2, 6, 7, 8, 9, 10, 12
biodiversity
    conservation in Bolivia, 174
    conservation in Cameroon, 190
    conservation in Kilum–Ijim Forest, 190–1, 193, 194, 202
    in Kavango, Namibia, 210
    in Maasai, 46, 50
    in Tonga, 125
BirdLife International, 190, 192, 193, 196, 203, 203n
Bolivia, 11
    biodiversity conservation in, 174
    deforestation in, 173, 177
    Law of Popular Participation 1996, 174
    Organizaciones Territoriales de Base (OTBs), 174
    soil conservation in, 171–88
Brazil, 224
Buckingham Canal, 65
Burkina Faso, 100

Cameroon, 11
    biodiversity conservation in, 190
    Community Forestry Management Unit, 195
    democratisation in, 196
    Forestry Law 1994, 194, 197, 203n
    Fulani, 191, 199
    Government of, 193
    Highlands Endemic Bird Area (EBA), 190, 203n
    Kilum–Ijim Forest, 12, 189–203

Kilum–Ijim Forest Project (KIFP), 189–203, 203n
    Ministry of Environment and Forestry (MINEF) of, 192–6, 200, 201
    Yaounde, 195
capability
    definition of, 204
    institutional, 204, 224
    research and development, 204–25
Centre for Research on New International Order (CRiNIEO), 64
CGIAR, 173
Chakraborty, R. N., 1, 2, 4, 6, 7, 11, 12, 13, 14
Chambers, R., 6
Club du Sahel, 104
Communal Areas Management for Indigenous Resources (CAMPFIRE), 5, 13, 113–28
    assumptions of, 119
    and indigenous knowledge, 122–7
    microcosmic versus macrocosmic view, 117
    and spirituality, 125–6, 127
community forestry
    CFM, 190, 193, 197, 200
    and equity, 129–49
    and ethnicity, 140–1, 144–5
    and gender, 141–2, 144–5
    history of, 131
    and income distribution, 137–40, 142–4
    and inter-group equity, 130, 142–5
    and intra-group equity, 130, 137–42
    in Nepal, 129–49
    and redistribution, 146
    and representation, 147
community-based
    conservation, 49
    institutions, 1
    wildlife reserve, 51

conservation
  and livelihood issues, 63–74
co-production, 8
Costa Rica, 221
Cousins, B., 33

Danida, 105
Department for International
  Development (DfID), 2, 64, 188,
  195, 203*n*, 205
desertification, 50
Devereux, S., 79
Dryland Applied Research and
  Extension Project (DAREP),
  216–19, 224*n*
Dyson-Hudson, 24

Economic and Social Research
  Council, 15*n*
environmental education, 39–60
Ethiopia, 25, 99
European Union, 46, 80, 82

FAO, 1, 181
fish farming, 10, 14, 73
forest(s)
  afforestation, 199
  agroforestry, 50, 80, 81, 85, 87,
    89–90, 119
  alternative monitoring methods of,
    106
  and CAMPFIRE, 118
  endangered species in, 2
  incentives for conservation of,
    198–9
  and local management in Sahel,
    99–112
  monitoring of, 10, 105
  reforestation, 13, 100
  regeneration, 2, 108, 201
  'scientific' versus 'laissez-faire'
    management of, 108–10, 111
  sustainable yields, 99–112
  and tourism, 2

Global Positioning Systems (GPS), 9
Goldsmith, A. A., 223
Grameen Bank, 224
Gujarat, 168

Hasler, R., 119
Hillman, E., 58
Hooper, P.R., 162

IIED, 104–5, 201
India, 2, 13
  All-India Coordinated Maize
    Improvement Project, 211–12,
    216
  Coastal Regulation Zone (CRZ), 73
  conservation projects, 63
  Department of Fisheries, 66
  Forest Department, 5
  Government of, 165
  Himalayas, 214–16
  JFM in, 5
  maize research in, 211–14, 220
  Ministry of Surface Transport, 72
  Ministry of Water Resources, 165
  National Water Policy 1987, 165
  participatory forest management in,
    13
  Supreme Court of, 73
  Tamil Nadu, 63–74
  Uttar Pradesh, 212, 222
  India Mark II Handpump, 155
Indian Agricultural Research Council,
  212
Indian Space Research Organisation,
  71
Indo-German Watershed
  Development Programme, 159
International Centre for Maize and
  Wheat Improvement (CIMMYT),
  213, 216
International Crop Research Institute
  for the Semi Arid Tropics
  (ICRISAT), 211
International Development Research
  Centre, 150, 156, 160, 166,
  168*n*
International Labour Organisation
  (ILO), 217, 218

Kabiri, N., 2, 6, 9, 11, 14
Karamoja, 19
  Alternative Dispute Resolution
    (ADR) in, 31–4, 35–8
  and droughts, 19–20

Karamoja – *continued*
 Karamoja Development Agency
  (KDA), 25
 Karamoja Task Force, 32
 traditional conflict management in,
  29–31, 35–8
 and tribal warfare, 19, 21–2, 24–5,
  28
Karimajong, 6, 9, 10
 definition of, 20
 and Elders' Councils, 33, 36
 and livestock mobility, 20
 local institutions, 23–6
Kelly, M., 1, 2, 4, 11, 12
Kenya, 19, 25, 26, 32
 Amboseli National Park, 42–3, 53
 British colonialism in, 40
 Kenya Wildlife Service, 43, 46, 47,
  50, 54
 Lake Amboseli, 43
 Ministry of Agriculture of, 217
 Mount Kilimanjaro, 42, 45
 Pokot, 27
 technology development in,
  216–18, 222
 Turkana Project, 102, 109
Kerkhof, P., 1, 2, 4, 6, 7, 10, 11, 12,
  13, 14
knowledge
 cosmovision, 5, 120, 121, 122–7
 and empowerment, 166
 and endogenous development, 127
 exchange, 165
 farming, 119–20, 171
 gender division of, 180
 and *in situ* learning, 215–6
 indigenous, 5, 7, 9, 13, 35, 119–20,
  122, 126–8, 157, 158, 160–1,
  171–88, 211
 local, 5, 9, 76
 romanticism of, 125
 and spirituality, 120, 121, 127, 157,
  165
 technology transfer, 171, 201
 transformation of, 171
 vocational training, 157–8, 167

land husbandry, 75–96
 benefits of, 77

and fertility, 80, 83, 85, 93, 94, 95,
  176–8
and organic matter, 176–8
and overgrazing, 173, 177, 178,
  182, 184
regulations, 79
soil and water conservation, 80, 83,
  94–5, 159, 165–7, 171, 179*t*,
  180–2, 183, 186–8, 216, 217
and soil erosion, 171, 174, 176–8
time-scale, 85
versus soil conservation, 76
Lawrence, A., 1, 2, 6, 7, 9, 11, 12, 13
Long, N., 171

Maasai, 6, 9, 11, 40–3
 agriculture, 43–5, 46, 47
 and biological diversity, 46, 50
 and conflict resolution, 49–50
 distrust of government, 49, 53,
  58–9
 economy, 41
 fencing, 46–7
 land alienation, 52, 56
 land use, 43–5
 and leadership struggles, 55
 and livestock–wildlife conflict, 48,
  50, 53
 salination, 45–6
 sedentrisation, 45
 spirituality, 51–2
 tourism, 46, 54, 56–67
 violent conflict, 57–8
Maharashtra
 Akole Taluka, 150–68
 Bharatiya Agro Industries
  Foundation (BAIF), 150, 152,
  155–9, 160–3, 166–8
 deforestation in, 154
 Government of, 165
 health problems in, 154, 155, 160,
  164, 166
 Mahadeo Kali tribe, 154
 Mahars caste, 154
 Pravara river, 152, 155
 Wadi (Orchard) Programme, 155
Malawi, 11, 75–96
 Agricultural Development Division
  (ADD), 81

Malawi – *continued*
  Banda, Dr. Hastings Kamuzu, 77
  Catchment Area Development
      Committee (CADC), 83–4, 94
  Catchment Area Development Plan
      (CADP)
  Government of, 79, 82
  Lake Malawi, 76–7
  land degradation in, 77
  Ministry of Agriculture of, 82, 83
  Ngodzi river, 78
  Salima Rural Development Project,
      82
  sustainable technologies, 83, 84
  United Democratic Front (UDF), 77
Mali, 100
  BICOF, 107
  local forestry rights, 106–7
  Rural Woodfuel Market, 101–4,
      107–8
  traditional institutions, 107
Martin, A., 2, 6, 7, 12
Mathew, S., 67
Matsaert, H., 2, 6, 7, 12
Maveneke, T.N., 116
Mengistu, 25
Mexico, 214

Namibia
  agricultural research in, 205, 206,
      207
  German immigrants in, 208
  Horticultural Research Unit, 208
  Kavango Farming Systems Research
      and Extension (KFSR/E), 205–7,
      209–11, 220, 222, 224*n*
  Kavango Seed Fair, 210, 219, 221
  Ministry of Agriculture, 205, 206,
      207
National Bank for Agriculture and
    Rural Development (NABARD),
    159
natural resource management
    (NRM), 1
  by centralised bureaucracies,
      189–90
  and devolution, 189–203
  CAMPFIRE, 113–28, 190
  community forestry, 129

  community-based, 19, 39, 43, 63
  and co-operative systems, 11
  and education, 50–1, 78, 93
  and gender issues, 105, 156, 164,
      184, 199
  and grazing, 51
  in Cameroon, 189–203
  indigenous knowledge, 119, 126–8,
      157, 171–88
  open access versus communal
      property, 116
  soil, 2, 75–96, 171–88
  stakeholder representation, 55
  water, 2, 150–68
  wildlife conservation, 116–17
Nepal, 7, 11, 13, 129–49
  District Forest Officer, 132, 143,
      147
  Forest Act 1993, 131, 132
  Forest Department, 131
  Forest Regulation 1995, 132
  Private Forest Nationalisation Act
      1957, 131
Netherlands Government, 203*n*
New Institutional Economics, 14–15
Niamir-Fuller, M., 2, 6, 9, 10, 11, 12,
    14
Niger
  constraints to land management,
      101
  Gestion de Terroir, 99–112
  livestock, 101–2
  Rural Code, 110
  Rural Woodfuel Market, 99–112
Norad, 109

OECD, 104
organic agricultural cooperative, 174,
    180, 181
Ostrom, E., 8, 13

Panini, D., 1, 2, 4, 6, 10, 11
Pantnagar University, 212, 213, 214
participation, 15
  and conservation, 53–4
  and forest management, 99
  and institutions, 12
  and land conservation, 76
  and land husbandry, 75–96

participation – *continued*
  as means or ends, 84
  multi-stakeholder constraints in,
    105
  and NRM, 12–13
  as orthodoxy, 1–3
  participatory rural appraisal, 2, 12,
    83, 173, 176, 181, 185
  and planning and development, 38
  and rolling back the state, 7
  and rural development, 40
  and soil conservation, 171–88
  Spanish meaning of, 186
  and technology development, 186
  typology of, 3, 82
  and water resources, 150–68
Pretty, J., 39, 82
Promotion of Soil Conservation and
    Rural Production (PROSCARP),
    76, 78–9, 83–6
  and agroforestry, 89–90, 94
  description of, 80–2
  Field Assistant (FA), 82, 87
  firewood, 90–1
  and health and nutrition, 91, 93, 94
  intervention areas of, 86
  participation statistics, 86
  and sanitation, 91–2, 94
  strategies, 81
  water, 92–3, 94
Pulicat Lake, 6, 10, 14, 63, 64
  description and ecology, 65–6
  effects of Ennore port, 72
  endangered species, 64
  fisherfolk and scientific project 64,
    70, 71
  fishing gear, 67–8, 70
  padu system 66–9
  salinity levels, 73
  scientific project and *padu*, 69–70
  scientific techniques of restoration,
    64
  shortcomings of *padu*, 70–1
  threats to *padu*, 71
  Wildlife (Protection) Act, 65

rapid rural appraisal (RRA), 155, 166
Richard, W., 58
Roling, N., 222

Rural Woodfuel Market
  in Burkina Faso, 100
  criticism of, 101
  in Niger, 100
  in Senegal, 100

Sahel
  conflict management in, 104
  constraints to local management,
    109–10
  forest ecology, 100
  forest management in, 99–112
  forest usage, 101–2
  Guesselbodi project, Niger, 100
  land use conflicts in, 103–5
  nomads in, 2, 101–5, 108, 110, 111
  and TRDP, Kenya, 100
Senegal, 100
Siade Barre, 25
Sibanda, B., 1, 2, 5, 6, 9, 11, 13, 14
Simpson, F., 1, 2, 6, 9, 14
social capital, 8, 9, 12, 29, 34–8,
    204–25
  and coalitions, 219, 222
  definition of, 204
  in Namibia, 206, 207, 208
social context analysis, 40, 52, 59
Sohani, G., 1, 2, 6, 9, 14
Somalia, 25
SOS Sahel, 10, 102, 103, 104, 106,
    107
South Africa, 2
South America, 2
Sri Lanka, 67
Subbarao, K.V., 162
Sudan, 19, 25, 27, 38, 40
  Forest Act 1989, 103, 108, 110
  registered forests, 108

Tanzania, 40, 99
Thomas, D., 1, 2, 4, 11, 12, 14
Tonga, 6, 9, 13, 113–28
  biological diversity, 125
  cosmovision, 120–2
  spirituality, 113–28
  tragedy of the commons, 27, 129
triticale, 214–16, 219–20, 221
Tropical Agricultural Research Centre
    (CIAT), 173, 174, 176, 180–6

Uganda
  Amin, I., 12, 22
  British Administration, 24, 26, 27,
    31, 38*n*
  Constitution of 1995, 28
  land tenure systems, 28
  Museveni, 12, 19, 24, 25
  National Parks, 27
  National Resistance Movement, 25
  Obote, M., 12
  Okello, 22
  Revolutionary Councils (RC), 24, 26
University of Waterloo, 10, 101, 162
University of Windsor (UW), 150,
    155, 156, 158, 162

Villareal, M., 171

Western D., 48
World Bank, 1

Energie II (Niger), 100
Environment Department, 2
Global Environment Facility, 203*n*
good governance, 7
Lilongwe Land Development
    Programme (Malawi), 78
Rural Woodfuel Markets, 107, 109
World Wide Fund for Nature, 203*n*
World Wide Fund India, 63, 64, 74
World Wide Fund UK, 64

Zambia, 114
Zimbabwe, 11, 113, 114
  Department of National Parks and
    Wild Life Management
    (DNPWLM), 116, 117, 126
  Kariba Dam, 114
  Ministry of Environment, 117
  National Parks and Wildlife Act
    1975, 116–17